Joanne O'Brien has worked for more than twenty years as a writer, researcher, and commentator on religious issues. She also heads the CIRCA RELIGION Photo Library.

Martin Palmer heads the International Consultancy of Religion, Education and Culture (ICOREC) and the Alliance of Religions and Conservation (ARC). He is also a broadcaster for the BBC.

Their other books include *Religions of the World* and *Festivals of the World.*

In the same series:

The Atlas of Climate Change
by Kirstin Dow
and Thomas E. Downing

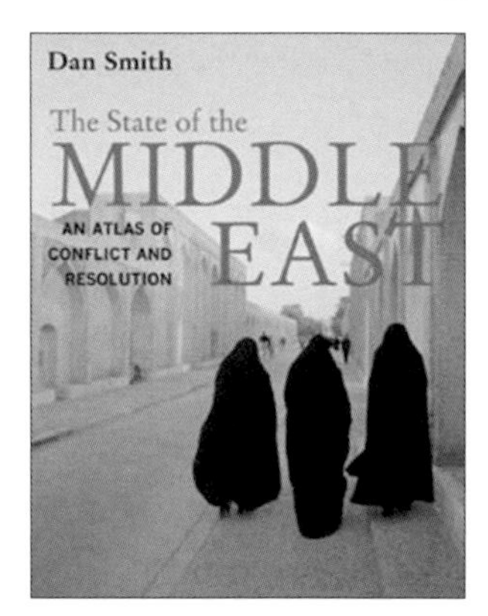

The State of the Middle East
by Dan Smith

The State of China Atlas
by Stephanie Hemelryk Donald
and Robert Benewick

THE ATLAS OF RELIGION

Joanne O'Brien and Martin Palmer

Consultant Editor
David B. Barrett

UNIVERSITY OF CALIFORNIA PRESS

Berkeley   Los Angeles

University of California Press, one of the most distinguished university presses in the United States, enriches lives around the world by advancing scholarship in the humanities, social sciences, and natural sciences. Its activities are supported by the UC Press Foundation and by philanthropic contributions from individuals and institutions. For more information, visit www.ucpress.edu.

University of California Press
Berkeley and Los Angeles, California

Cataloging-in-publication data for this title
is on file with the Library of Congress.

ISBN-13: 978-0-520-24917-2

Produced for the University of California Press by
Myriad Editions Limited
Brighton, UK
www.MyriadEditions.com

Directed by Candida Lacey
Edited by Jannet King and Candida Lacey
with Sadie Mayne
Maps created by Isabelle Lewis
Design and graphics by
Isabelle Lewis and Corinne Pearlman
Proofread by Elizabeth Wyse

Printed on paper produced from sustainable sources.
Printed and bound in Hong Kong through Phoenix Offset Limited
under the supervision of Bob Cassels, The Hanway Press, London.

15 14 13 12 11 10 09 08 07
10 9 8 7 6 5 4 3 2 1

Contents

Acknowledgements

Special thanks to Hannah Welton who assisted with the research and checking on a number of maps, and to the team at Myriad – editors Jannet King, Candida Lacey and Sadie Mayne, and designers Isabelle Lewis and Corinne Pearlman – for their encouragement and dedicated work. We wish to extend our thanks also to Stephen Robinson for advice and help with the statistical compilation of the data, to Victoria Finlay for comments and insights, and to many colleagues in ARC whose reports, thoughts and reflections over the last couple of years have helped to shape our understanding of the world and its faiths, in particular, Tara Lewis, John Smith, Mike Shackleton, Guido Verboom, Paola Triolo and Dr He Xiaoxin. Above all, our gratitude to David B Barrett, whose knowledge, experience, detailed research and generosity of spirit have made this book possible.

Joanne O'Brien
Martin Palmer

Photo credits

The publishers are grateful to the following for permission to reproduce their photographs:

CIRCA RELIGION Photo Library www.circalibrary.com
12: Lotus, William Holtby; 20: Hindu guru, Bip Mistry; 22–23: March for Jesus, Washington DC, Martin Palmer; Orthodox Christians, Mike Edwards; 24–25: Mosque in Pakistan, William Holtby; Muslims at prayer, William Holtby; 26–27: Ganesha, Bip Mistry; Brother and sister, Bip Mistry; Shrine, Bip Mistry; 28–29: Nuns in pink robes, William Holtby; Buddhist monks, William Holtby; 30–31: Prayer at Western Wall, Barrie Searle; Torah scrolls, Mike Edwards; 32–33: Guru Nanak, Twin Studio; Reading the Guru Granth Sahib, John Smith; 34–35: Siberian Shaman, Maxim Shaposhnikov; Tayakh, Maxim Shaposhnikov; 36–37: Stained glass window, John Fryer; 44: Muslim women, William Holtby; 54–55: Open Bible, Mike Edwards; Painted Ethiopian Bible, John Smith; 56–57: Chinese Church, Tjalling Halbertsma; Turkey, John Smith; 60–61: Iran, William Holtby; 62: Jerusalem soldiers, Zbigniew Kosc; 64–65: Jerusalem gate, Zbigniew Kosc; 68: March for Jesus, Martin Palmer; 74–75: Brazil, Benedictine Community of Serra Clara, Itajuba; Cambodia, Association of Buddhism for the Environment, Cambodia; 76–77: Pakistan Market, William Holtby; 80: Hanging Temple, Tjalling Halbertsma; 82–83: Guru Nanak, Twin Studio; Lao Zi, Tjalling Halbertsma; Bodhidharma, Martin Palmer; 86–87: Mount Tai Shan, Tjalling Halbertsma; 90: Buddhist monk, Bip Mistry

www.iStockphoto.com
34–35: Inukshuk, Shaun Lowe/iStockphoto; 60–61: Grand Mosque, Casablanca, Morocco, Michel de Nijs/iStockphoto; Djenne Mosque, Mali, Bytestrolch/iStockphoto; 82–83: Salt Lake City Temple, Utah, USA, Peter Chen/iStockphoto; St Peter's, Rome, Italy, Ricardo Garza/iStockphoto; St Sofia's Cathedral, Kiev, Ukraine, Vassili Koretski/iStockphoto; Golden Temple, Kyoto, Japan, pixonaut/iStockphoto; 86–87: Kootenay National Park, BC, Canada, Ulrike Hammerich/iStockphoto; Wupatki National Monument, Arizona, USA, Jason Cheever/iStockphoto; 87: Meteora Rousanou, Greece, Ben van der Zee/iStockphoto; St Catherine's Monastery, Sinai, Egypt, Vladimir Pomortsev/iStockphoto; Mount Kinabulu, Borneo, Malaysia, Wei Yee Koay/iStockphoto

Other sources:
70–71: Council for a Parliament of the World's Religions, Steve Rohrbach; 74–75: Mexico, Pro-Natura-Chiapas; Eco-coffins, Working for Water's Invasive Alien Species Programme, South Africa.

Introduction

It was a fond hope of many secular ideologies in the last 100 years that 'religion will wither and die' in the light of social and technological advancements. From marxism to fascism, the expectation was that religion would be thrown aside as an emotional and intellectual prop that was no longer needed.

This expectation was largely turned on its head during the last 20 years, and continues to be so. Instead, it has tended to be the ideologies that have faded, not the religions. In this new edition of the atlas, first published in 1993, we have a whole map that records this. 'Emerging from Persecution' (pp 66–67) charts the unprecedented levels of persecution against all major religions that has taken place since the beginning of the 20th century. More religious buildings have been damaged or destroyed, and more people have been killed for their religion during this period than in any previous century. Yet religious communities have recovered and are, in many, cases as numerous as before. Whether it is Daoists in China – almost wiped out during the Cultural Revolution of 1966 to 1974, the Tibetan Buddhists – persecuted by communism from 1949 to the present day, or Russian Orthodox Christians – attacked from 1917 until the 1980s, religion has not withered and in part, the need for an atlas such as this reflects that fact.

Religious communities have also, on occasion, proved more sustainable than nation states. For example, when Zambia gained independence in 1964, the schools, colleges and health networks of hospitals and clinics run by the churches and the mosques were nationalized. The failure of the Zambian state has meant that many of these, often in a state of physical and economic collapse, have been handed back to the churches and mosques to run again, as we show in the map 'Shared World' (pp 70–71).

The flip-side of the return or rise of the role of religion has been a growth in religiously inspired violence. While it is true to say that the majority of the last 100 years saw secular ideologies attacking religions, today that has been replaced by religions attacking secular societies, or waging war on other religious communities and identities. The recent rise of terrorists claiming to act in the name of Islam, the resurgence of religious conflict in a previously stable and multi-faith Indonesia, or the role of militant Buddhism in Sri Lanka's long running civil war underlines why we all need to know and understand more about the religious nature of the world today.

The re-emergence of old antagonisms, rooted in history, has led us to create another new map in this revised edition. The map on 'Faultlines' (pp 64–65) highlights the deep and ancient faultlines that run across parts of the world, and which have their origins in religious tensions. The violence of the 'civil war' in Iraq between Sunni and Shi'a has its roots in conflicts going back to the 7th century, and the opposition by much of 'Christian' Europe to the entry of 'Islamic' Turkey goes back to the 14th century.

The business of religion is also covered here. The dramatic growth in Islamic banking and in ethical investment movements within the religions is recorded – something the wider world has yet to truly awaken to. Likewise, the sheer scale of religious funding and support of those suffering from HIV/AIDS, and the level of development funding from Islam and Christianity are things the world rarely acknowledges.

The heart of the atlas remains the fascinating story of what the major religions in the world are, where they are today and how much power they have. The story of which religion is predominant in a given country is told both through numbers in the map 'Popular Religions' (pp 14–15) and in the three supporting maps, 'Arrivals' (pp 16–17), 'Roots and Branches' (pp 18–19) and 'Origins' (pp 82–83). These help to fill out a picture of the overall range and spread of faiths culturally, geographically and historically.

Supporting this are the maps of specific religions, which introduce us to a shifting and changing religious world – be that Sikhs in the USA, the recovery of Buddhism in Asia or the role and sheer size of Catholicism worldwide.

In trying to give a picture of the scale and nature of religious involvement in the world, we touch upon relationships with the state, the role of women, and the decline in many indigenous religions and the corresponding rise of newer forms of religious identity that draw heavily upon traditional and indigenous religions such as Shamanism. The adaptability of religion is one of its key features. Religions are able to convince us that they are unchanging, yet they survive and spread precisely because they are constantly adapting. We have tried to capture something of this dynamic solidity in the maps.

Take for example, the place of religions in the environment movement. In the first edition, we explored the rising role of religions as partners in environmental protection. This is recorded again, and illustrates a dramatic rise in the extent and level of such involvement. But alongside this we have the map 'Holy Natural' (pp 84–85), which demonstrates that many of the world's most significant national parks, wilderness areas and protected environments are also sacred sites, and that this sacredness has helped protect them for centuries, if not millennia.

Undertaking a task such as this has been tough, but we have been guided by one of the world's leading religious statisticians, Dr David B Barrett, whose encyclopedic knowledge and vast database have been invaluable, as have his insights into the significance of much of the data.

We have called upon the services of many researchers, from bodies as diverse as, for example, INFORM, the World Bank, WWF and specialists in religious education around the world, and owe a huge debt of gratitude to those who have given so much of their time and knowledge to make this as good a production as possible. Any inadequacies are the responsibility solely of the authors.

In the years since the first edition the internet has arisen – itself a tool of considerable significance to the workings of religion. For the first edition, we had to rely on hunting down books and articles. For this edition we have had to ensure that the thousands of fascinating websites we have visited had some authority – and alongside printed reference works we have also used these internet resources.

We believe that, as the 'Future' map (pp 78–79) indicates, the role of religion will continue to grow and to have increased influence upon other aspects of society. Some will view this with alarm, others with a sense of success. From our perspective, we can but note that the religions are the world's oldest human institutions. They have lasted for millennia because they understand what it is to be human and they know how to help us through the stages of life. Without them our world is a duller, greyer and less joyful place for many. With them we encounter not just the grace of religion but also at times the curse of religion. Religion is not going to wither and die. It is up to us to ensure that it contributes to a wider, pluralistic society rather than a narrower one. And this is a challenge that most religions welcome and embrace.

Joanne O'Brien
Martin Palmer

Part One BEGINNINGS

Religions arise in a diversity of ways. Some, such as Daoism, emerge from the lifestyles and beliefs, environment and landscape of the people. Some, such as Zoroastrianism, are believed to have been revealed. Yet others develop from the spiritual and philosophical experiences of the founder, such as happened in the case of Buddhism. Some remain rooted in a given location – such as Shinto in Japan. Others have spread around the world. The rise of missionary religions (starting with Buddhism in the 3rd century BCE), followed by Judaism, Christianity and Islam, are the best-known examples.

Some religions have only recently started to move beyond their homelands. Hinduism and Daoism are the best-known examples of this. Yet others have to all intents and purposes been reinvented – Shamanism, for example, is now a term used to describe a vast plethora of different practices around the world.

Within each major religion, differences of opinion, often over issues of authority and power, led to splits. These traditions themselves then frequently gave rise to further schisms. For example, Anglicanism split away from the Catholic Church and then Baptists and Methodists split from the Anglican Church.

Religions also spawn new religions, as Judaism has in relation to Christianity. Although Islam established itself as an independent religious tradition, for several centuries some Christians viewed it as a schism within Christianity. To this day, Jains sometimes find themselves viewed as Hindus by Hindu organizations. While occasionally religious traditions reunite, usually it is the other way round and the number and range of diverse traditions within each religion keeps growing.

While the religious maps of Europe and Asia have remained almost unchanged for centuries, those of the Americas, Africa and much of the Pacific have been radically transformed in the last 200 years, and are still in a state of flux. They bear witness to the trade winds, and to the rise of Europe as a spiritual as well as economic and military power. As that power wanes, newer branches of Christianity in particular are taking on an autonomy that is often challenging to the theologies and ideas of the older European Churches. Similar trends can be seen in other missionary religions, such as Islam, and in particular Buddhism as it spreads into the Western world.

Lotus flower, symbol of wisdom and clarity, arising out of the cloudy waters of ignorance

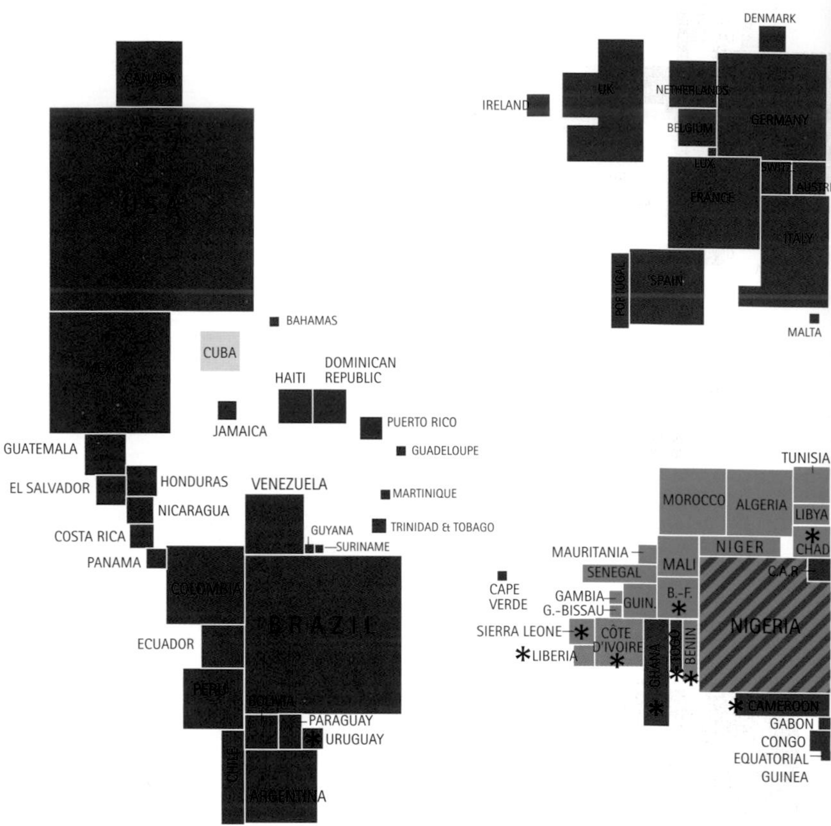

While 80 percent of people worldwide profess some religious allegiance, what this means differs from country to country and even from religion to religion.

In Islam, the notion that religion is separate from daily life is unthinkable: it is a way of life rather than a faith. Similarly, Hindus see what they believe as being how they live. There is no sense of one set of beliefs for everyday life and another for religious life. In fact, Hinduism as a term of reference to a 'religion' is an external creation: the name was introduced by the Persians to describe all beliefs in India – across the River Indus. Judaism is also particular, since it is both a way of life and an ethnic identity – though not always linked to religious belief or practice.

For many people, religious identity goes hand in hand with ethnic, social and cultural identity. Thus, questions about how much a religion is practised are not appropriate to Indonesia, for example, nor to large swathes of Africa, South America or even China. For many people, religion is not a choice. They are born into a given set of values and beliefs. Unless some major trauma shakes them or they move right away from their own culture, the religion of their birth remains lifelong. This pattern can be disturbed. Certain religions and new religious movements are committed to conversion, and the arrival of missionaries can change religious allegiance.

In areas where religions are expanding quickly – notably in Eastern Europe and Africa – religious commitment often carries with it powerful social, political and ethnic identity.

Popular Religions

Allegiance to a single religion is professed by at least two-thirds of the population in more than 80% of the world's states.

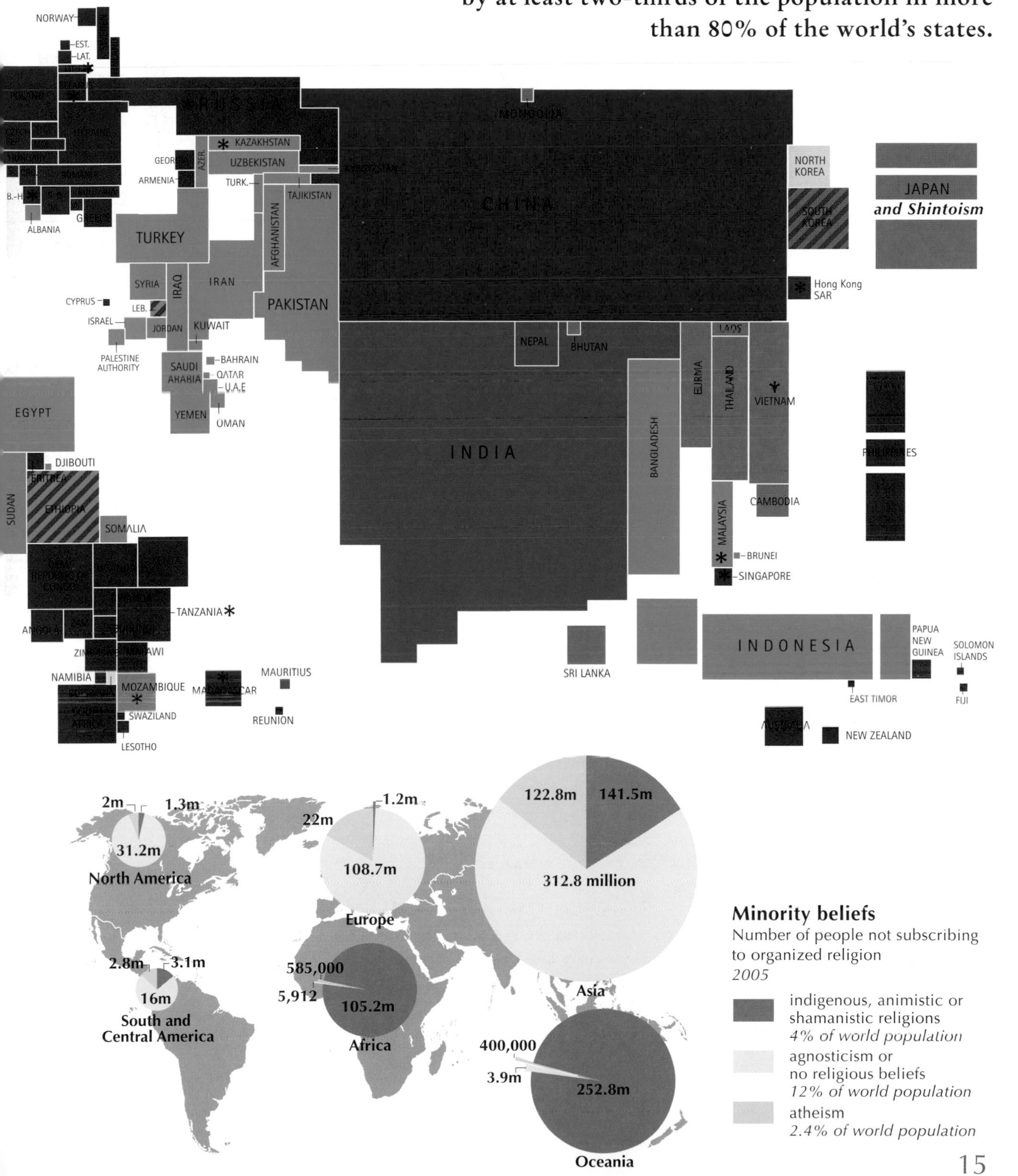

In the 3rd century BCE, the Emperor Ashoka, whose empire covered most of present-day India and Pakistan, sent out missionaries to spread the Buddhist teachings from Sri Lanka to Egypt. Prior to this, religions appear to have been ethnically or culturally based, with no principle of seeking to convert others to a different way of life.

The rise of the missionary ushered in a new world of international religions. By the 1st century BCE, Judaism had become a missionary religion, and from it sprang the two most successful missionary religions in history: Christianity in the 1st century CE and Islam in the 7th century.

The European trading nations brought about the next major change. From 1450 onwards, Portugal, Spain, England and the Netherlands followed new sea routes both westwards, to the Americas, and eastward to Asia, because Islam was blocking the old trade routes through the Middle East. This is why southern Africa, the Americas, Australasia and the Pacific islands are now largely Christian. Russia, similarly avoiding Islamic countries, expanded across Siberia and arrived north of China in the late 17th century.

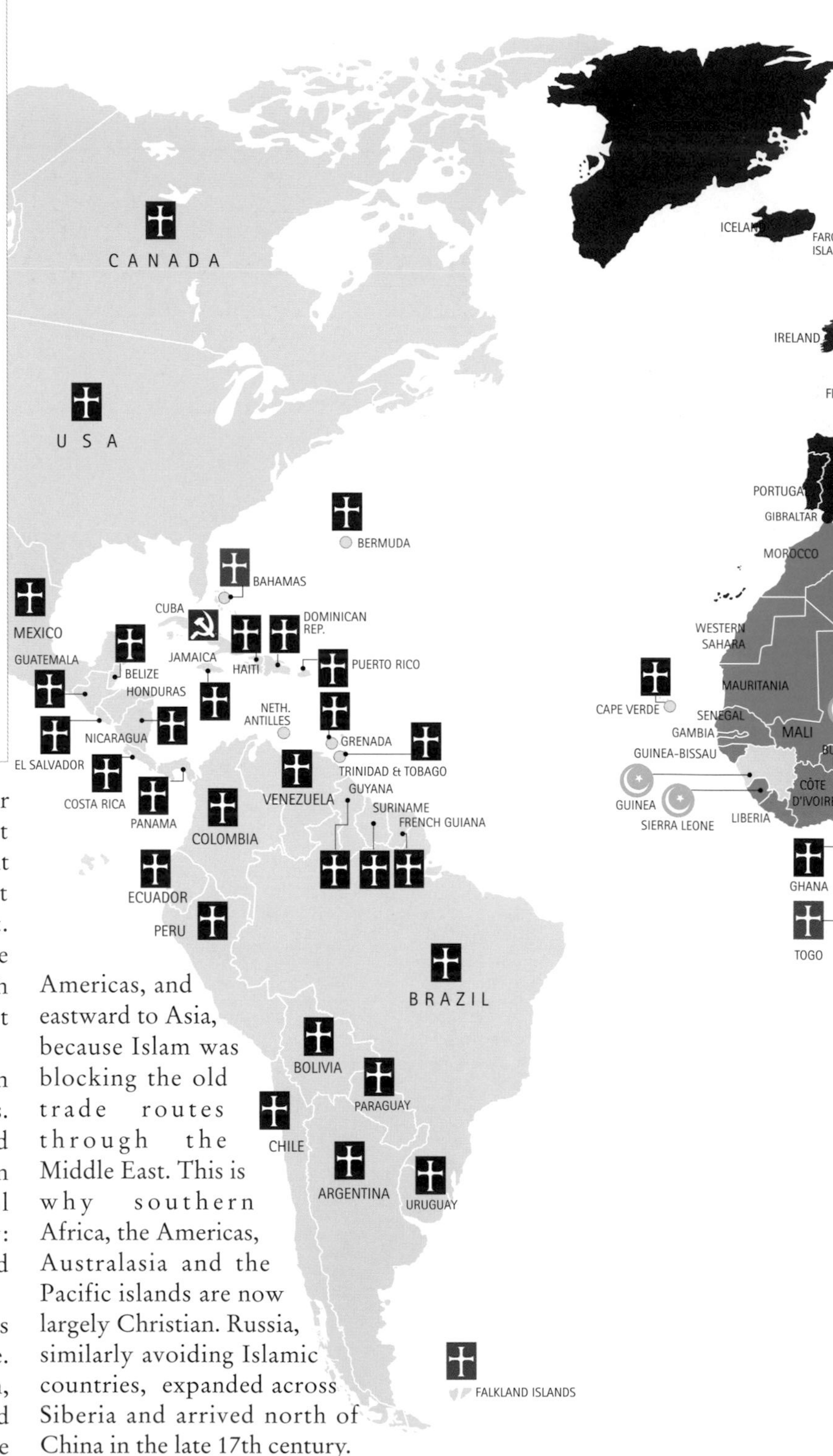

Arrivals

The rise of the missionary religions and seaborne trade has greatly influenced the religious map of the world.

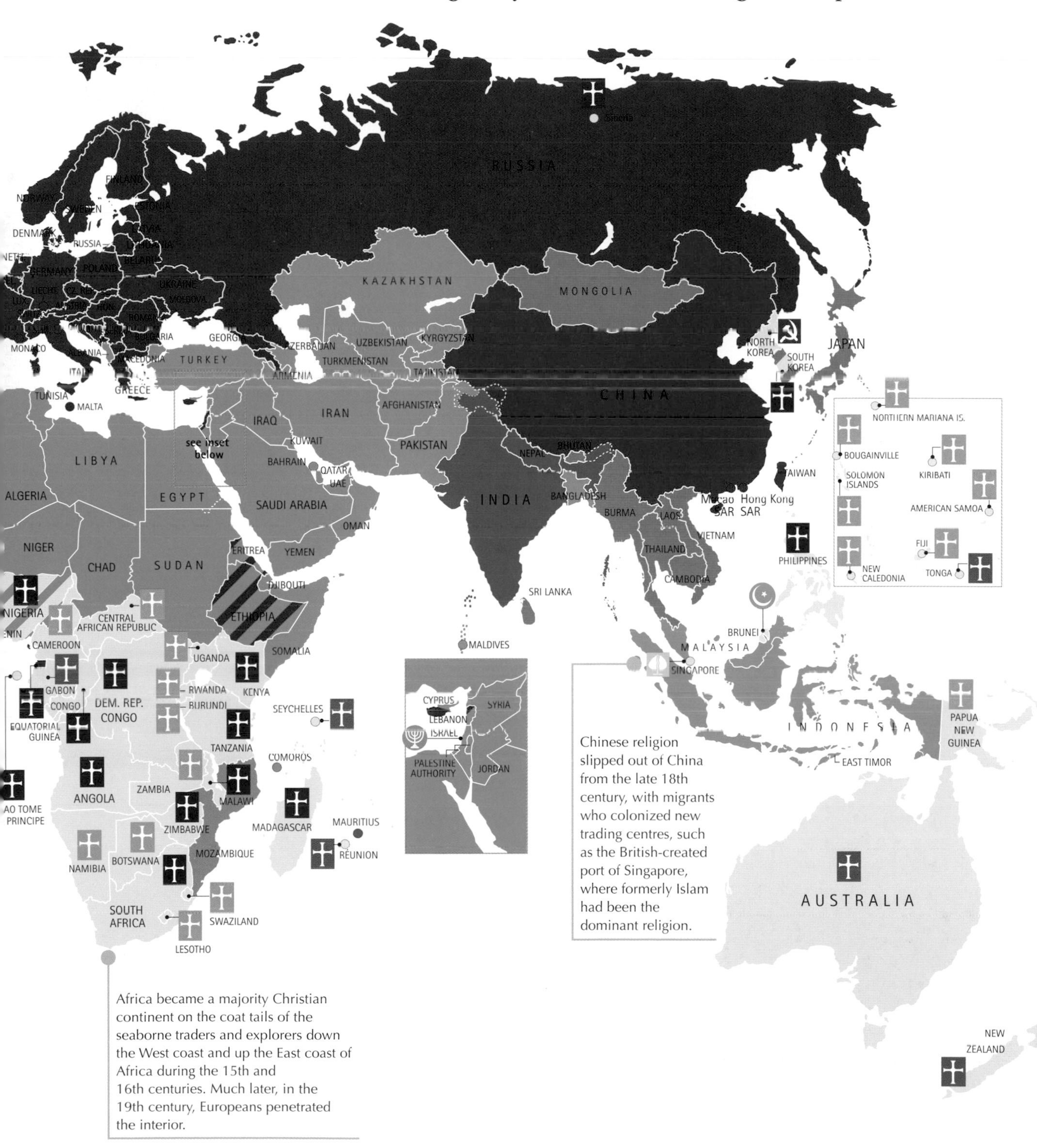

Chinese religion slipped out of China from the late 18th century, with migrants who colonized new trading centres, such as the British-created port of Singapore, where formerly Islam had been the dominant religion.

Africa became a majority Christian continent on the coat tails of the seaborne traders and explorers down the West coast and up the East coast of Africa during the 15th and 16th centuries. Much later, in the 19th century, Europeans penetrated the interior.

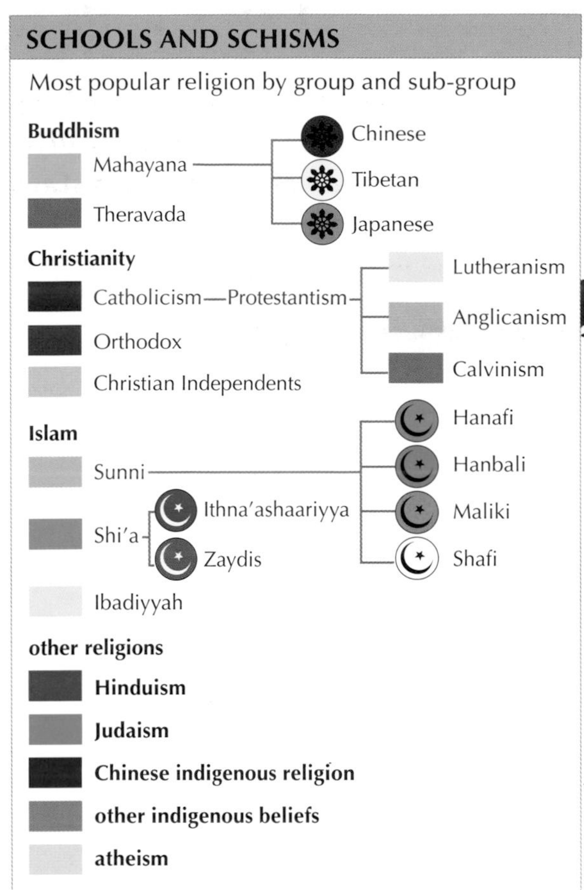

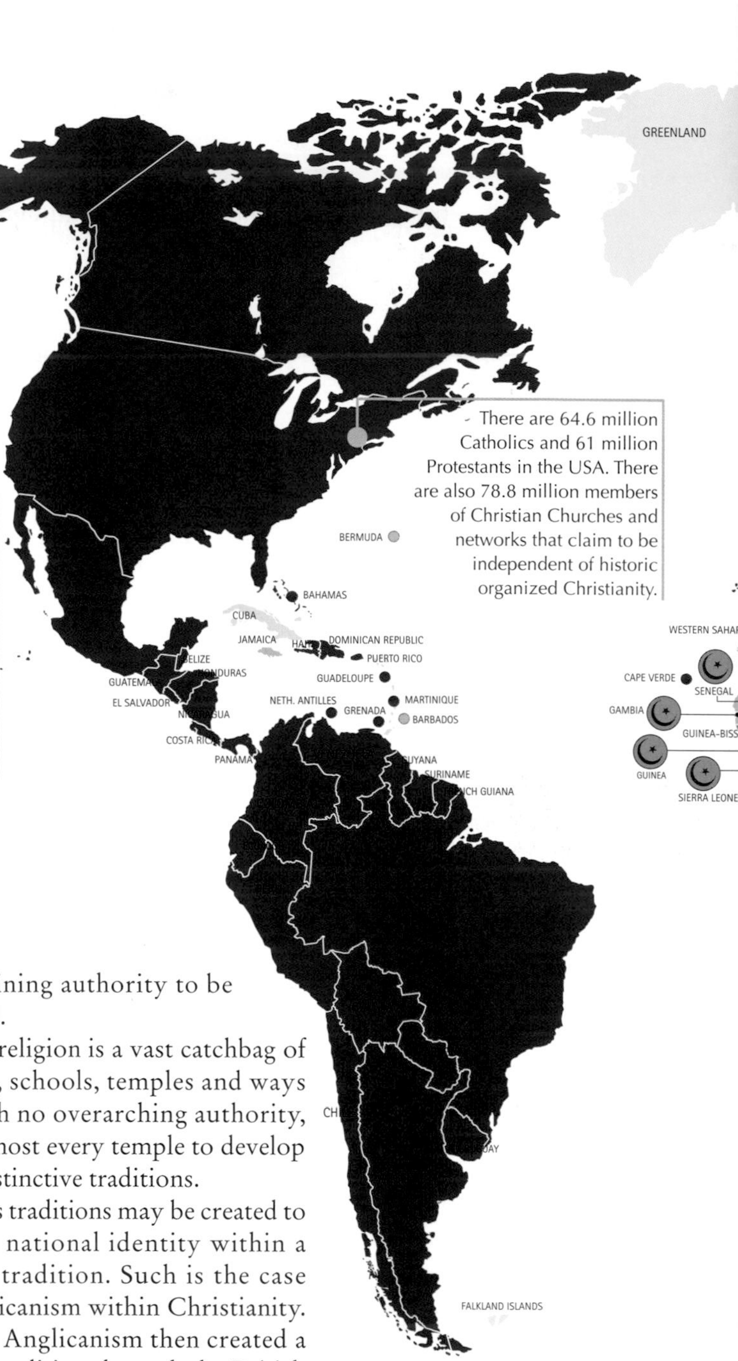

The different traditions within one religion owe their creation to many different factors. In the process of expanding beyond its original heartland, a religion can take on the beliefs of another culture – as happened when Buddhism moved from India through China to Japan. Political divisions may exacerbate theological differences and help to create alternative traditions – as occurred when the demise of the Roman Empire split the Christian Church in two: the Eastern Orthodox Church, based in Constantinople, and the Western Catholic Church, based in Rome.

Divisions often arise when a religious structure claims to be authoritative – as happened when the Protestants broke from the domination of Rome. Faiths with a more fluid organizational structure, such as Hinduism, tend to have fewer splits because there is no single defining authority to be challenged.

Chinese religion is a vast catchbag of traditions, schools, temples and ways of life with no overarching authority, leaving almost every temple to develop its own distinctive traditions.

Religious traditions may be created to support a national identity within a universal tradition. Such is the case with Anglicanism within Christianity. However, Anglicanism then created a universal tradition through the British Empire, and this is now splitting into new traditions as cultural factors, particularly in African traditions, challenge the earlier Anglo-American traditions.

Roots and Branches

Yesterday's revolution can be today's orthodoxy. Some new movements have grown to become the most popular religion in a state.

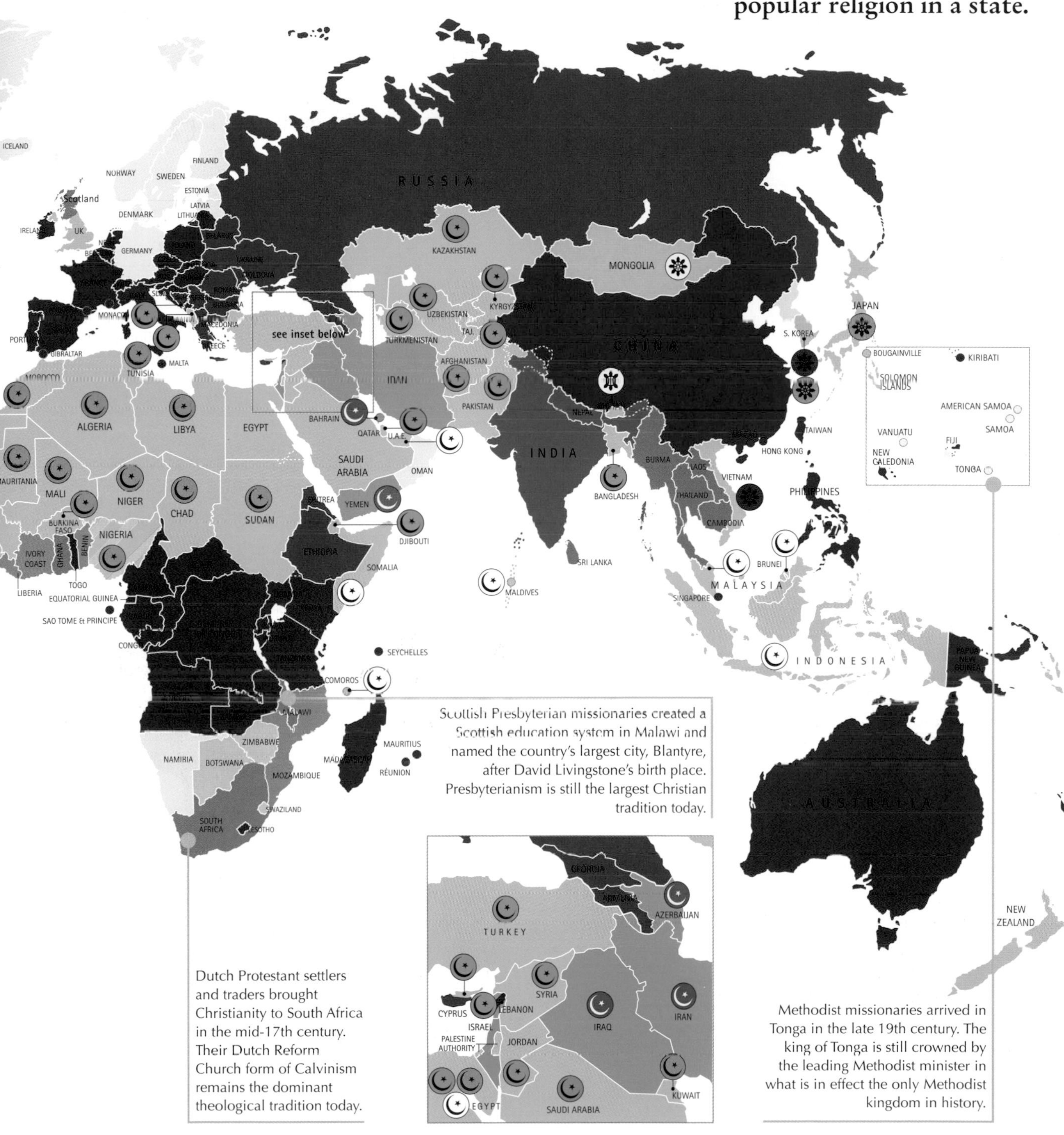

Scottish Presbyterian missionaries created a Scottish education system in Malawi and named the country's largest city, Blantyre, after David Livingstone's birth place. Presbyterianism is still the largest Christian tradition today.

Dutch Protestant settlers and traders brought Christianity to South Africa in the mid-17th century. Their Dutch Reform Church form of Calvinism remains the dominant theological tradition today.

Methodist missionaries arrived in Tonga in the late 19th century. The king of Tonga is still crowned by the leading Methodist minister in what is in effect the only Methodist kingdom in history.

Part Two BELIEFS

The vast majority of people follow one or other of the major world religions. While there is a small, but growing number of non-believers, the number of atheists has shrunk recently due to the fall of communism.

Adherence to one religion is not the only way that people express religious commitment. China, for example, has a complex mixture of religious traditions, ranging from Shamanism, through Daoism to Buddhism and Christianity. In practice, most Chinese make use of different aspects of each tradition for different needs. For example, they might use Shamanism or Daoism for exorcisms, Daoism for charms and magic, Buddhism for death ceremonies, and Christianity for success in business deals. Combining these into a workable mix is not considered a problem by most Chinese, whereas the Western and Islamic world view asks for adherence to just one tradition.

For many who have grown up within a specific religion, the pluralism of today offers additional elements of spirituality and religious practice that people are beginning to use increasingly, while still remaining within the fold of one particular tradition. Many devout Catholics, for example, will practise yoga, while devout Buddhists will also offer prayers to the Virgin Mary in times of special need. This wider framework of religious reference is also underpinning the growth in new religious movements and the surprising growth of indigenous traditions, such as Native American and traditional African religions, which have spread to many other parts of the world.

In looking at the figures for religious belief, it is useful to remember that outward adherence often covers a vast array of different beliefs, rooted in one tradition, but increasingly fed by many sources of inspiration.

A Hindu sadhu, Varanasi, India. The marks on his forehead denote that he is a Saiva, a follower of the god Shiva

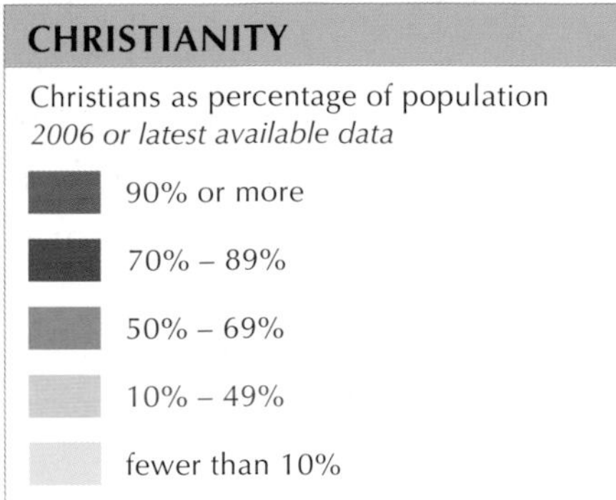

A third of all people belong to one of six major Christian traditions. Within these traditions there is wide variation, creating a vast array of interpretation and practice. There are over 33,800 Christian denominations in 238 countries, comprising 3.4 million worship centres, churches or congregations.

Christianity is growing across the world, especially in South Korea, Russia and Sub-Saharan Africa. While the percentage of the populations of India and China that are Christian is small, there are sizeable Christian groups in these countries, some of whom may be hidden.

One of the emerging trends in Christianity has been the growth of independent movements. There has always been an historical tradition of church members leaving a parent body because of differences regarding authority, structure or lifestyle, and forming their own Christian groups. In time, many of these have become mainstream denominations, as has happened with Lutheranism or Methodism. Today's 'Independents' follow that tradition in their search for a church lifestyle and authority appropriate to their beliefs, and often to their cultural and ethnic identity. Some denominations on the fringe of organized, mainstream Christianity are termed 'marginal Christians'.

Christianity

Christianity is the world's largest religion, with more than 2.1 billion adherents worldwide, and more than 33,000 denominations.

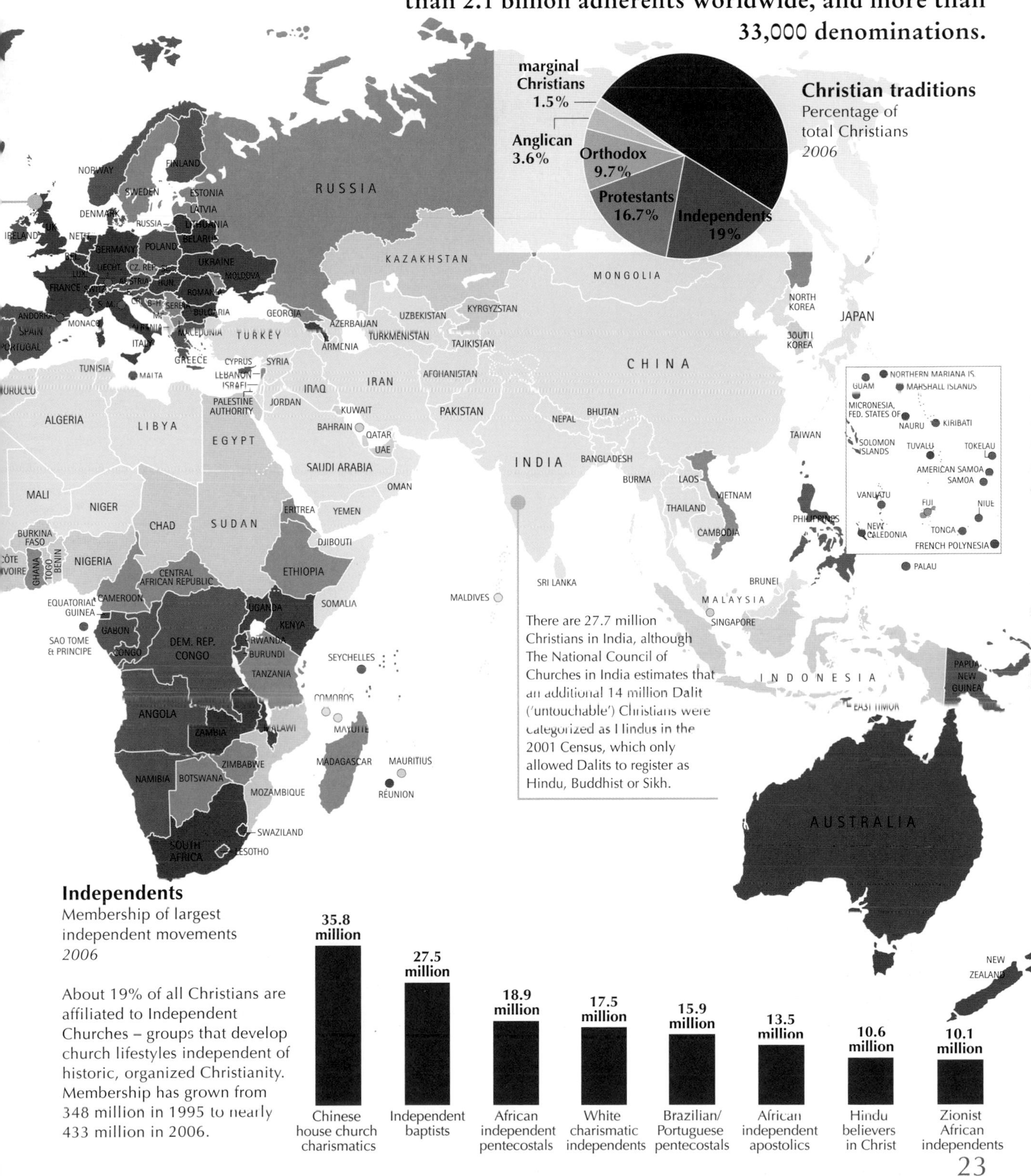

There are 27.7 million Christians in India, although The National Council of Churches in India estimates that an additional 14 million Dalit ('untouchable') Christians were categorized as Hindus in the 2001 Census, which only allowed Dalits to register as Hindu, Buddhist or Sikh.

Independents

Membership of largest independent movements
2006

About 19% of all Christians are affiliated to Independent Churches – groups that develop church lifestyles independent of historic, organized Christianity. Membership has grown from 348 million in 1995 to nearly 433 million in 2006.

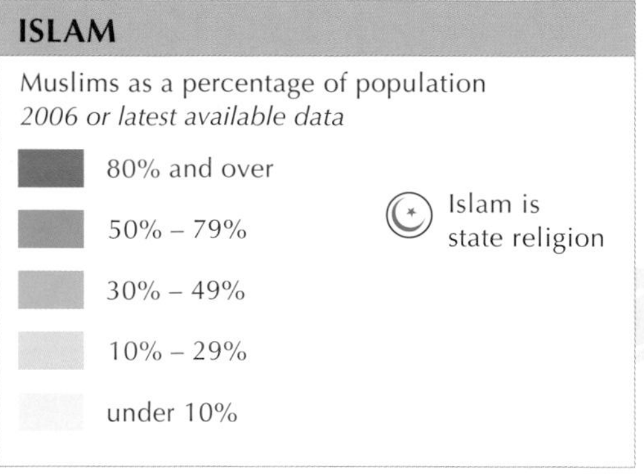

The two main traditions within Islam are Sunni and Shi'a. After the death of the Prophet Muhammad, leadership of the Muslim community passed to a succession of caliphs ('deputies'). In the mid-7th century under the caliphate of Ali, the Prophet Muhammad's son-in-law, some came to believe that leadership of the Muslim community should be hereditary; these became known as the Shi'a (partisans of Ali). The majority held that the caliphs should be democratically chosen, according to the Sunna, the sayings and customs of the Prophet Muhammad. These are known as Sunni. They number more than 1.1 billion, and are in the majority in most Islamic countries. The Shi'a number 192 million and are in the majority in Iran, Iraq, Bahrain, Azerbaijan and Yemen.

Within Sunni and Shi'a there are various schools of law and traditions. The Sunni schools of law are widespread, such as the Maliki school that is dominant throughout most of Muslim Africa. In Shi'a Islam there are more localized traditions, such as the Alawite and Druze in Syria and Lebanon.

The Ibadiyyah tradition originated in the decades immediately after the death of the Prophet Muhammad, before the split between Sunni and Shi'a. Small numbers of Ibadites, who are predominantly Bedouin Arabs, are found in the deserts of Arabia, Iraq and North Africa.

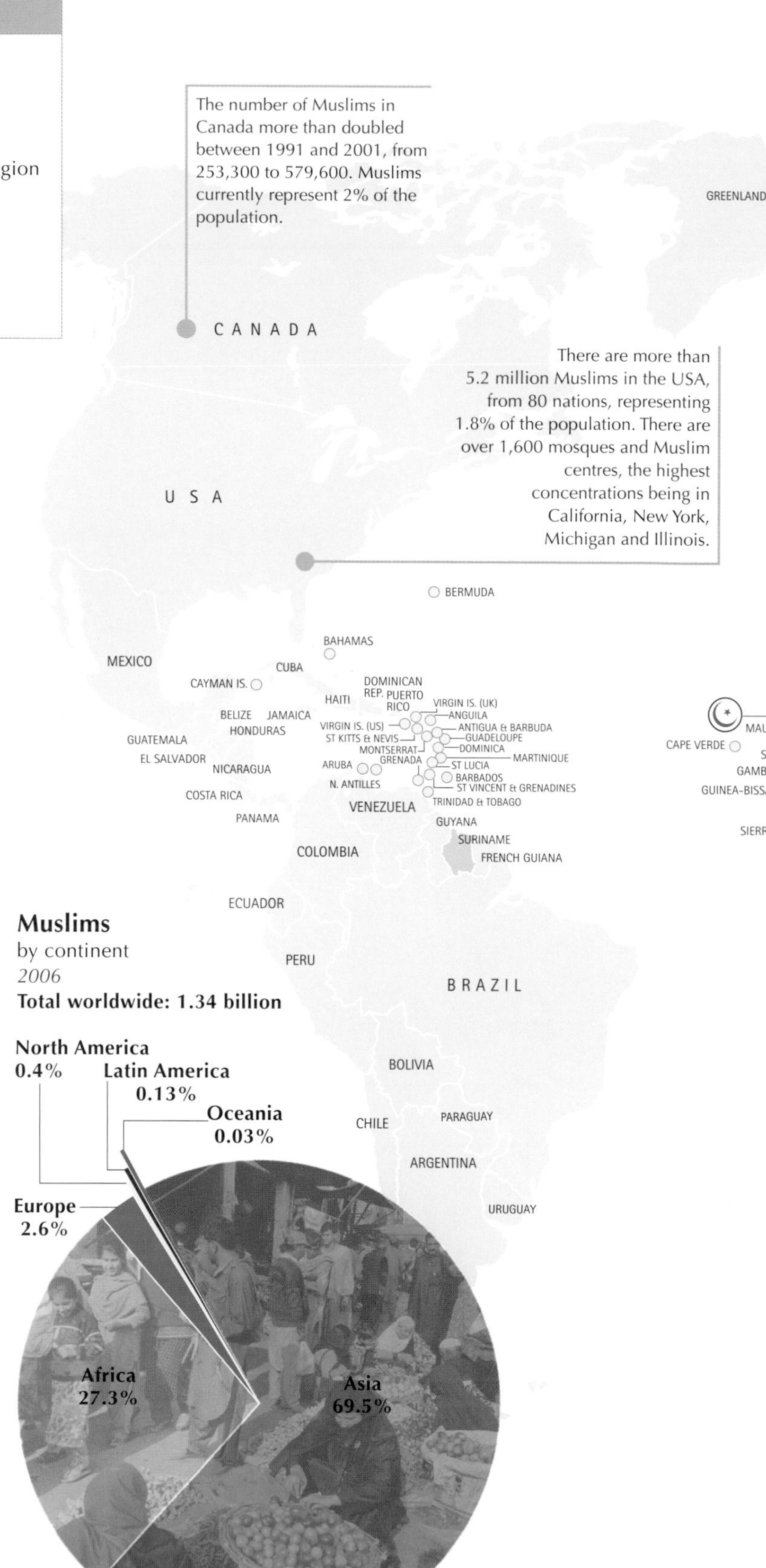

Islam

There are 1.34 billion Muslims worldwide, 20% of the world's population. Islam is the state religion of 25 countries.

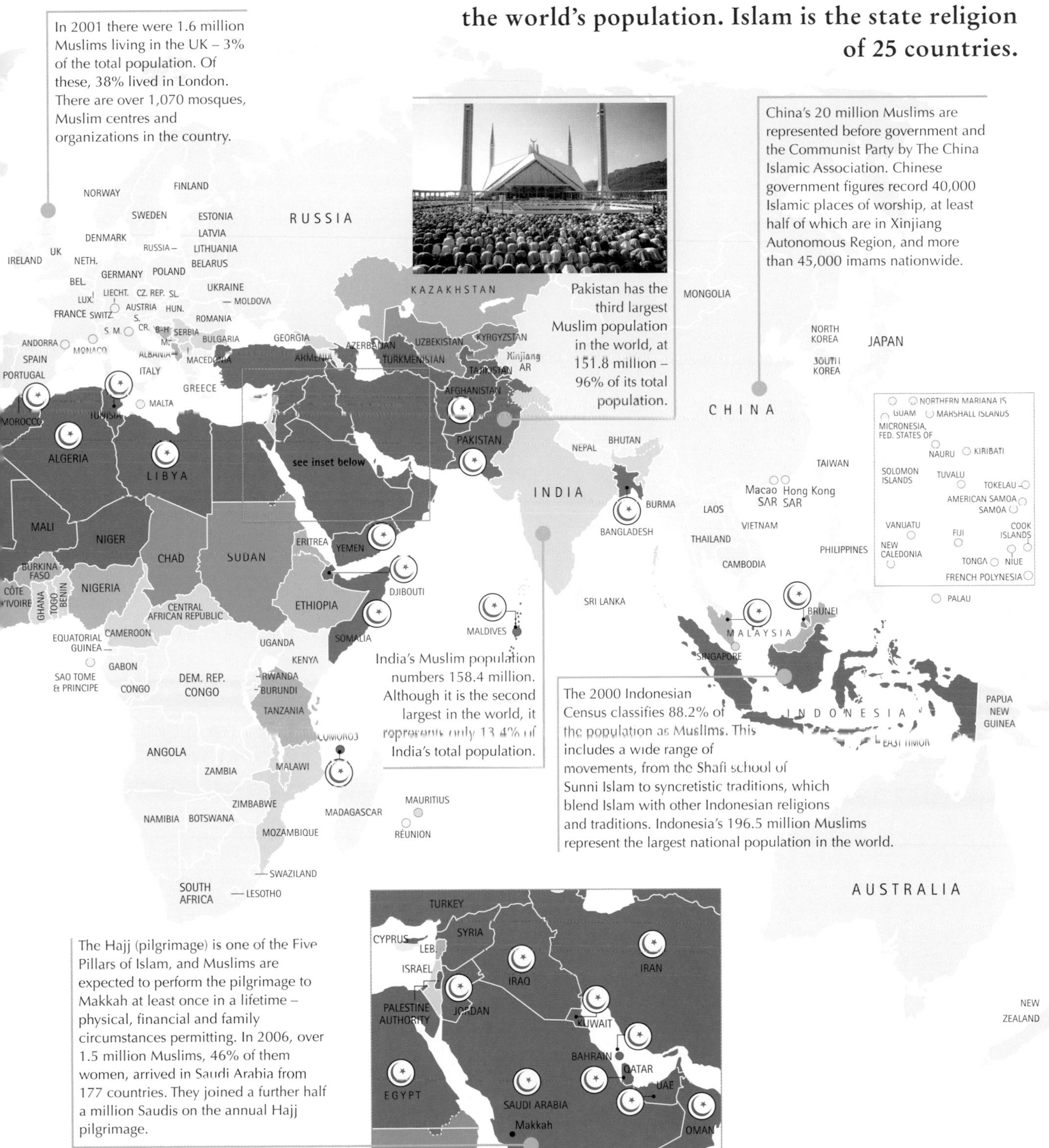

In 2001 there were 1.6 million Muslims living in the UK – 3% of the total population. Of these, 38% lived in London. There are over 1,070 mosques, Muslim centres and organizations in the country.

Pakistan has the third largest Muslim population in the world, at 151.8 million – 96% of its total population.

China's 20 million Muslims are represented before government and the Communist Party by The China Islamic Association. Chinese government figures record 40,000 Islamic places of worship, at least half of which are in Xinjiang Autonomous Region, and more than 45,000 imams nationwide.

India's Muslim population numbers 158.4 million. Although it is the second largest in the world, it represents only 13.4% of India's total population.

The 2000 Indonesian Census classifies 88.2% of the population as Muslims. This includes a wide range of movements, from the Shafi school of Sunni Islam to syncretistic traditions, which blend Islam with other Indonesian religions and traditions. Indonesia's 196.5 million Muslims represent the largest national population in the world.

The Hajj (pilgrimage) is one of the Five Pillars of Islam, and Muslims are expected to perform the pilgrimage to Makkah at least once in a lifetime – physical, financial and family circumstances permitting. In 2006, over 1.5 million Muslims, 46% of them women, arrived in Saudi Arabia from 177 countries. They joined a further half a million Saudis on the annual Hajj pilgrimage.

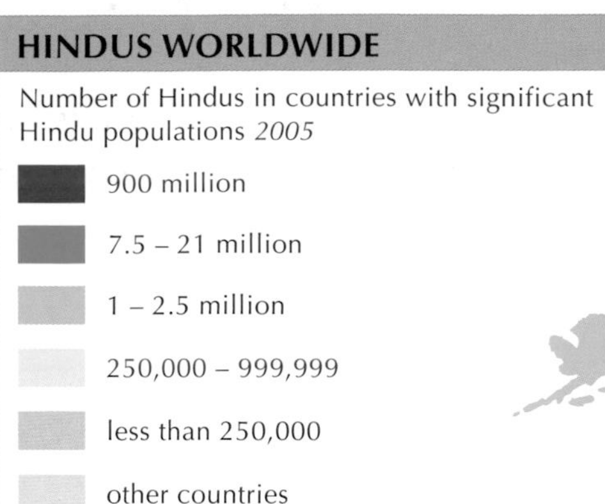

HINDUS WORLDWIDE

Number of Hindus in countries with significant Hindu populations *2005*

- 900 million
- 7.5 – 21 million
- 1 – 2.5 million
- 250,000 – 999,999
- less than 250,000
- other countries

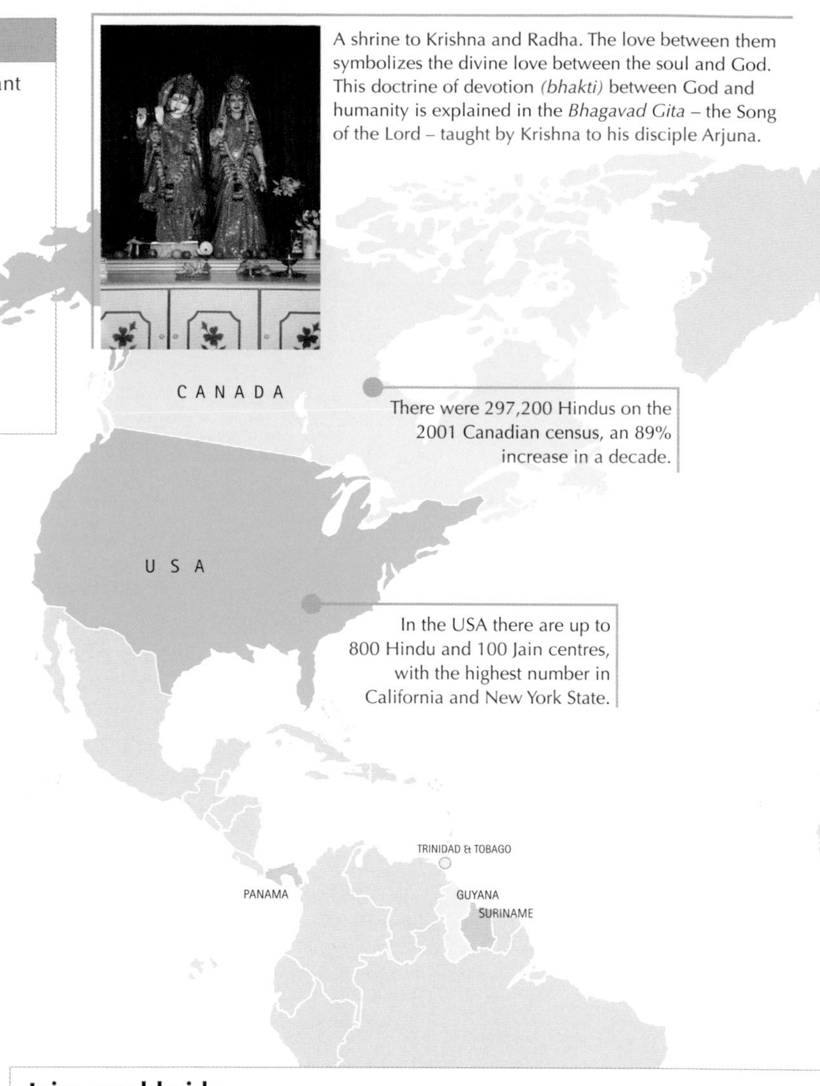

A shrine to Krishna and Radha. The love between them symbolizes the divine love between the soul and God. This doctrine of devotion *(bhakti)* between God and humanity is explained in the *Bhagavad Gita* – the Song of the Lord – taught by Krishna to his disciple Arjuna.

There were 297,200 Hindus on the 2001 Canadian census, an 89% increase in a decade.

In the USA there are up to 800 Hindu and 100 Jain centres, with the highest number in California and New York State.

Hinduism is intricately woven into the land and culture of India, and Hindus refer to their religion as *sanatana dharma*, the eternal truth or ancient religion.

Although 95 percent of the world's Hindus still live in India, there was a significant wave of migration from the 1st to 7th centuries CE along trade routes into South-East Asia and Indonesia. The second major phase began when Indians migrated to other parts of the British Empire as indented labour or for trade.

Many Hindu gurus travelled west from the late 19th century, and the spread of Hindu ideas has been considerable through Europe, North America and Australasia. In the USA, Swami Vivekananda spoke at the World's Parliament of Religions in 1893 in Chicago, and in 1894 established the Vedanta Society of New York. The Vendanta Society is also credited with establishing the first Hindu temple in the USA – in San Francisco in 1906.

Since the 1960s, a wave of movements inspired by Hindu philosophy and spirituality has emerged as westerners have been influenced by the teachings of visiting gurus, travels to India and translations of classics such as the *Bhagavad Gita* or the *Upanishads*. Movements such as the International Society for Krishna Consciousness (ISKCON) now have communities worldwide.

Jains worldwide

Countries with 2,000 or more Jains *2005*

Jains worldwide: 4.6 million

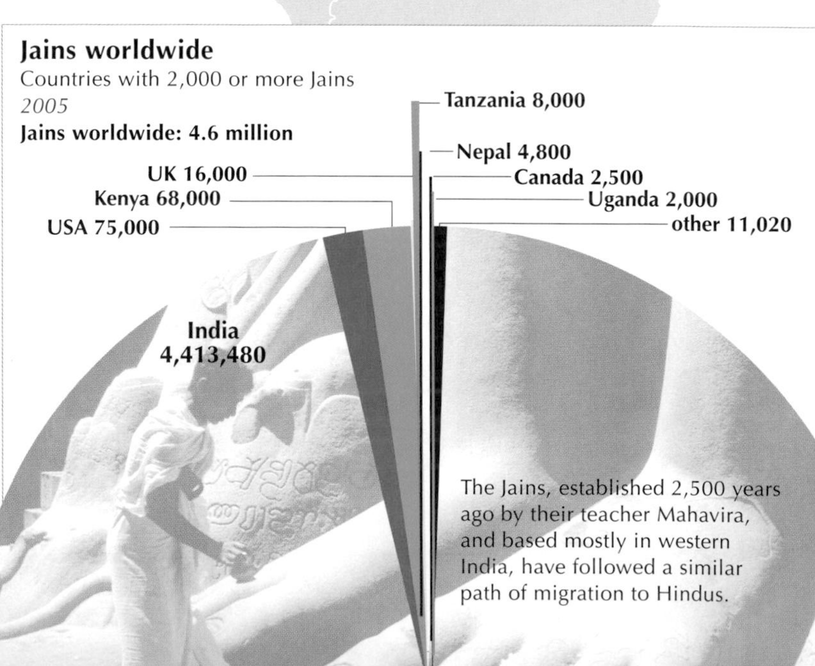

The Jains, established 2,500 years ago by their teacher Mahavira, and based mostly in western India, have followed a similar path of migration to Hindus.

Hinduism

Hinduism is the world's third largest religion, with over 950 million Hindus worldwide. Almost all live in South Asia, with the majority in India, where over 80% of people are Hindu.

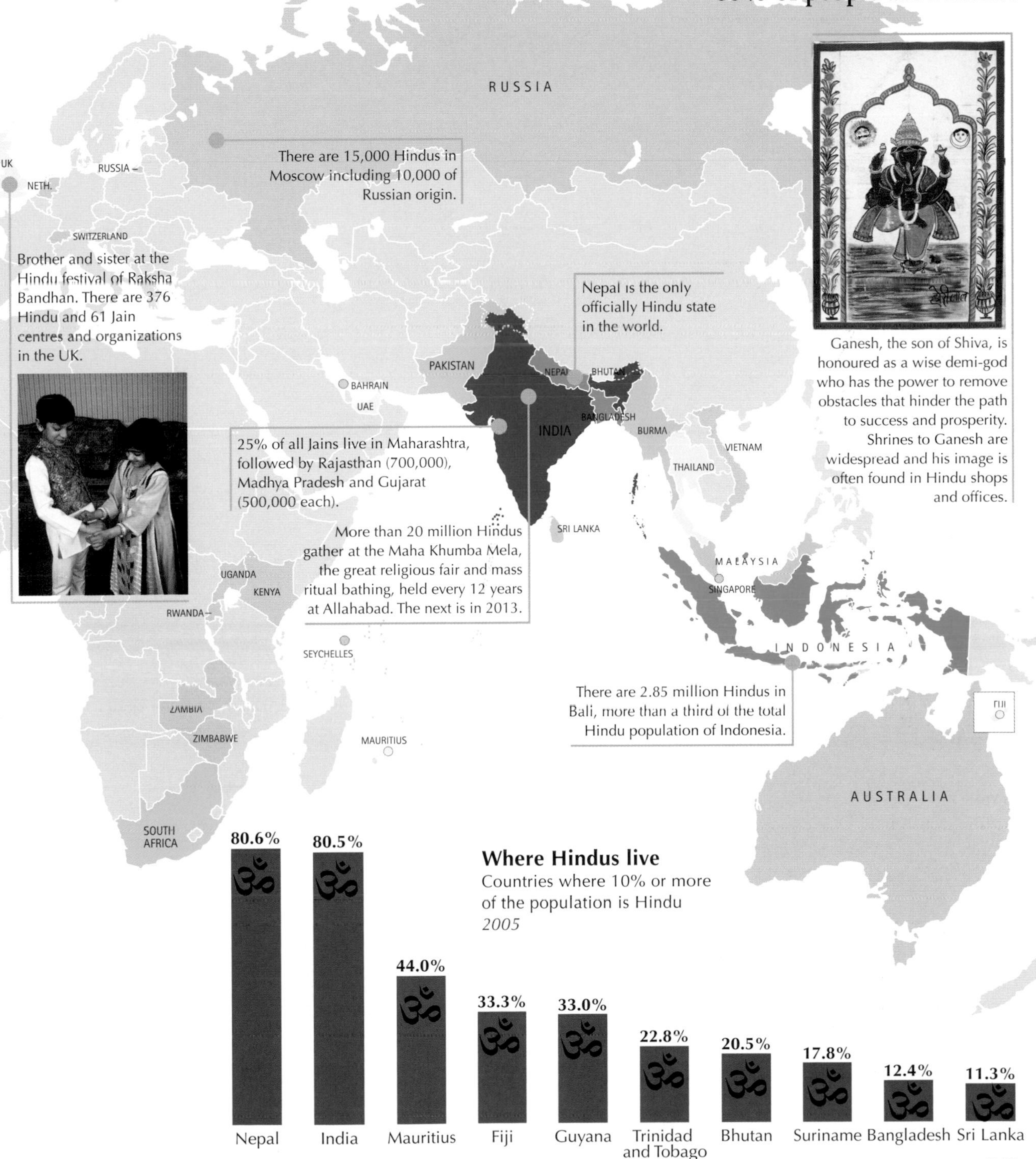

BUDDHISM IN ASIA

Countries with significant Buddhist populations
2006 or latest available data

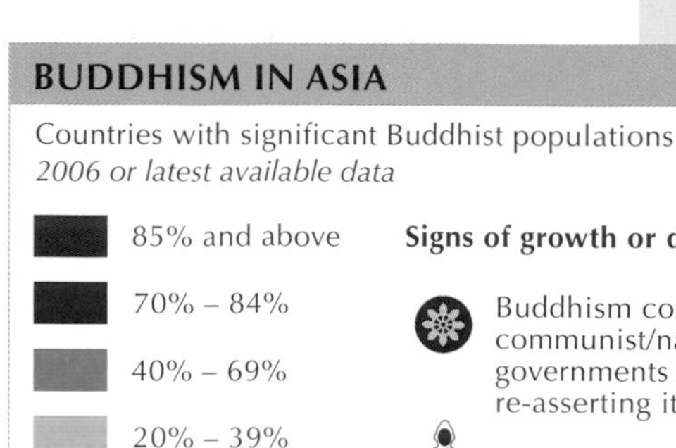

85% and above
70% – 84%
40% – 69%
20% – 39%
10% – 19%
below 10%
some Buddhist centres
other countries

Signs of growth or decline

Buddhism controlled by communist/nationalist governments but re-asserting itself

Buddhist monks and nuns

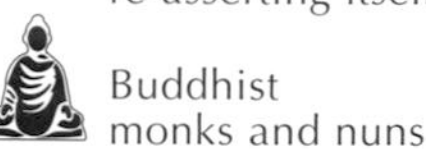

Buddhist temples or monasteries

declining Buddhist minorities

new Buddhist movements

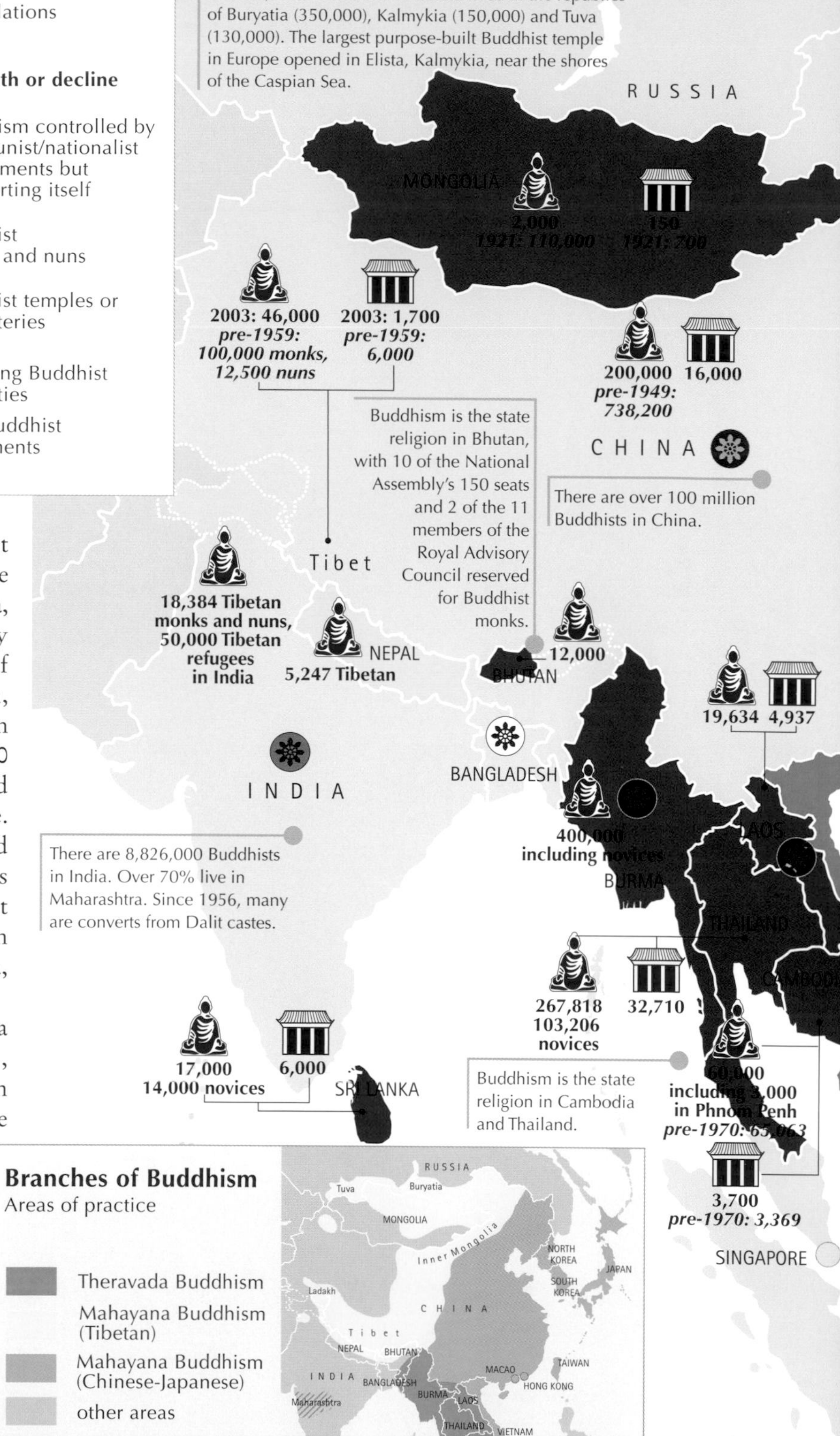

Buddhists make up nearly 6 percent of the world's population. More than 98 percent live in South-East Asia, where new political freedoms in many countries have witnessed a growth of Buddhist practice and monasticism, particularly since the 1990s. In Cambodia, for example, only 3,000 monks were thought to have survived the civil war and its ensuing genocide. By 2006 the number of monks and pagodas had returned to pre-war levels of 1969–70. However, Buddhist activities are monitored or restricted in other countries, such as North Korea, China and Burma.

Outside Asia, there has been a considerable growth in Buddhism, with more than 3 million Buddhists in the USA, over 1.5 million in Europe and almost 700,000 in Latin America.

Within the three major traditions or branches of Buddhism there are hundreds of smaller organizations and groups, including the English Sangha Trust, with 10,000 supporters; the Dharma Realm Buddhist Association, with 800,000 members, and Pure Land Buddhism in Japan, with 19.5 million members.

Buddhism

More than half the world's population live in countries where Buddhism is now, or has been, dominant. During the 20th century, Buddhism was subject to greater suppression than at any time in its history.

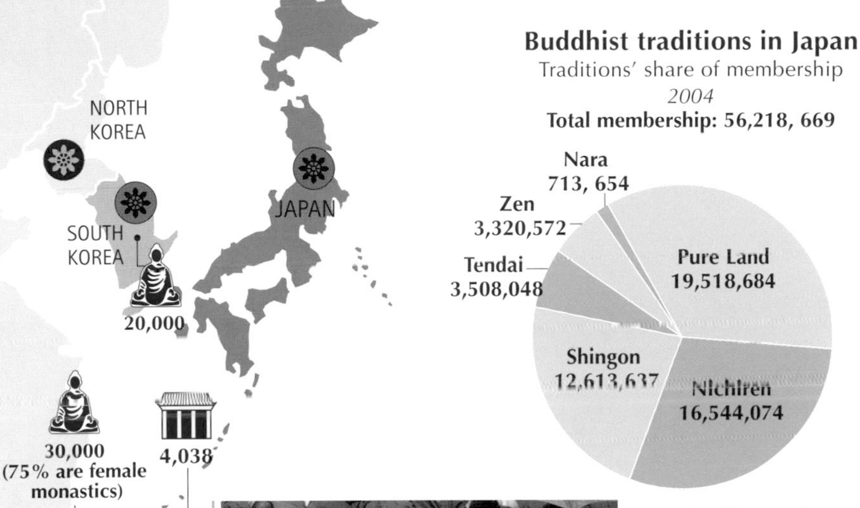

Nuns offering devotion at a Buddhist shrine in Burma. Although the military government of Burma is not anti-Buddhist, it monitors Buddhist activities, and friction exists between monastic communities and the government.

Tibetan Buddhist monks at Dharamsala in India, gathered to hear the Dalai Lama teaching. After the Chinese invasion of Tibet in 1951 and the suppression of the national uprising in Lhasa by Chinese troops in 1959, the Dalai Lama established a Tibetan 'government in exile' at Dharamsala.

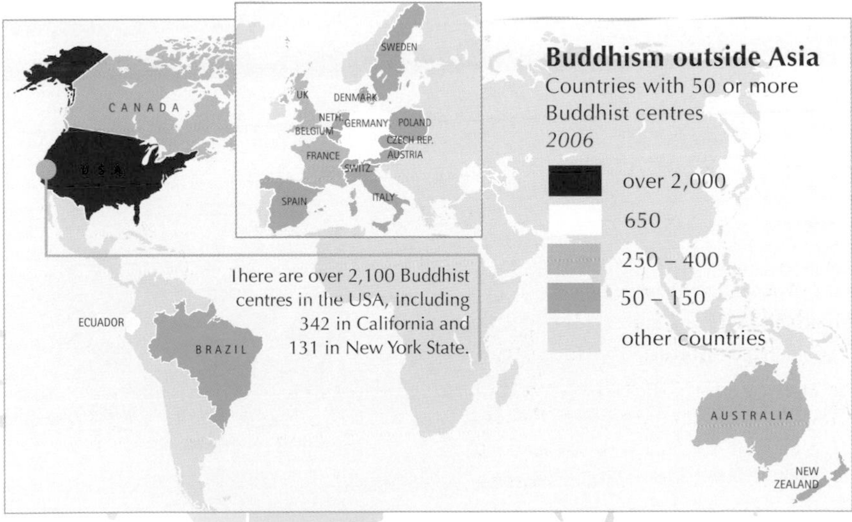

The scrolls of the Torah held aloft during Barmitzvah celebrations next to the Western Wall, the single remaining wall of the Second Temple in Jerusalem destroyed in 70 CE.

Since the 1980s, Jewish population trends have been affected by major socioeconomic and geopolitical changes: the break-up of the Soviet Union, the end of apartheid in South Africa, the reunification of Germany, the expansion of the European Union, instability in some Latin American countries and continuing tensions in the Middle East. As a result, 80 percent of the world's Jews now live in the USA and Israel.

The 'core' Jewish populations illustrated on this map include those who, when asked, identified themselves as Jews, or were identified by someone in the same household as Jews. This definition reflects subjective feelings and overlaps, but does not always coincide with, definitions of rabbinic law – Halakah. Jewish population studies also recognize the 'enlarged' Jewish population: this includes the 'core' population, all others of Jewish parentage who are not currently Jewish, and the non-Jewish members of their households.

In the state of Israel, however, individual status is subject to Ministry of Interior rulings that follow criteria established by rabbinical authorities. The 'core' Jewish population in Israel is not based on personal subjective identification but reflects the legal rules of Halakah.

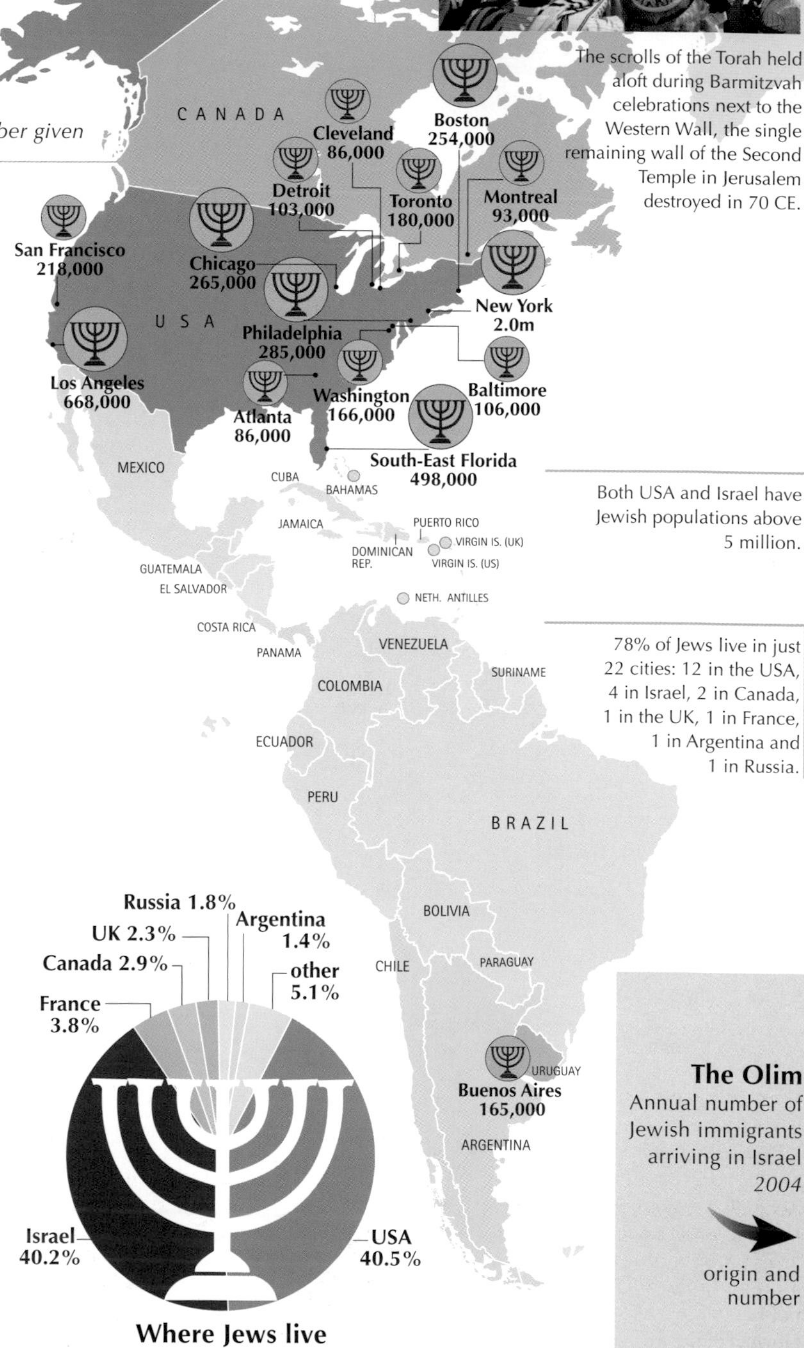

Both USA and Israel have Jewish populations above 5 million.

78% of Jews live in just 22 cities: 12 in the USA, 4 in Israel, 2 in Canada, 1 in the UK, 1 in France, 1 in Argentina and 1 in Russia.

Where Jews live
Countries' share of the world's Jewish population
2005 percentages

The Olim
Annual number of Jewish immigrants arriving in Israel
2004
origin and number

Judaism

There are over 13 million Jews worldwide, more than 5 million of whom live in Israel.

Nearly 3 million Jews have migrated to Israel since its foundation in 1948. Since 1989, over a million have come from the former Soviet Union.

Torah scrolls wrapped in a mantle. It is believed that God revealed the first five books of the Hebrew Bible, known collectively as the Torah or by the Greek term the Pentateuch, directly to Moses.

In 1950 Israel's Knesset passed a law beginning with the words 'Every Jew has the right to immigrate to his country'. According to the Law of Return in Israel, a Jew is any person born to a Jewish mother or who has converted to any denomination of Judaism and does not have another religious identity. However, the law currently extends provision to all current Jews, their children, grandchildren and respective Jewish or non-Jewish spouses.

The Sikh faith began in the Punjab region of India in the 15th century under the teachings of Guru Nanak. The Punjab remains the heartland of the Sikh religion, with up to 16 million Sikhs living in this state.

The largest Sikh populations outside India are in the USA, UK and Canada, and were originally established through links with British rule in India during the British Empire. Canada's first gurdwara opened in 1908 in Vancouver, the UK's in London in 1911, and the USA's in 1912 in California.

The 2001 British Census asked the population to voluntarily state their religion and 336,179 Sikhs in the UK were recorded. The Canadian Census of the same year asked for religious affiliation and recorded 278,400 Sikhs. In contrast, the US Bureau of Census is prohibited by law from asking this question. However, an independent Religions Congregations Membership Study (RCMS) is planning a major survey of religious populations in 2010 that will include a county-by-county statistic of the Sikh population in the USA. This development is welcomed by the World Sikh Council's America Region, which estimates that around 500,000 Sikhs are living in the USA.

A Sikh reading The Guru Granth Sahib in the House of God, also known as the Golden Temple, in Amritsar, Punjab. Sikh scripture, written in Gurmukhi, is contained in the Guru Granth Sahib, which is regarded as the eleventh and final Guru in the Sikh faith.

Sikhism

There are 24 million Sikhs worldwide. Over 90% live in India, mainly in the Punjab.

Guru Nanak (1469-1539) taught a new understanding of God from which Sikhism grew. Regarded as the first of the 11 Sikh gurus, he established communities in many parts of India, and travelled as far as Sri Lanka and Afghanistan. His teachings underlined the equality of all, service to the community and devotion to God.

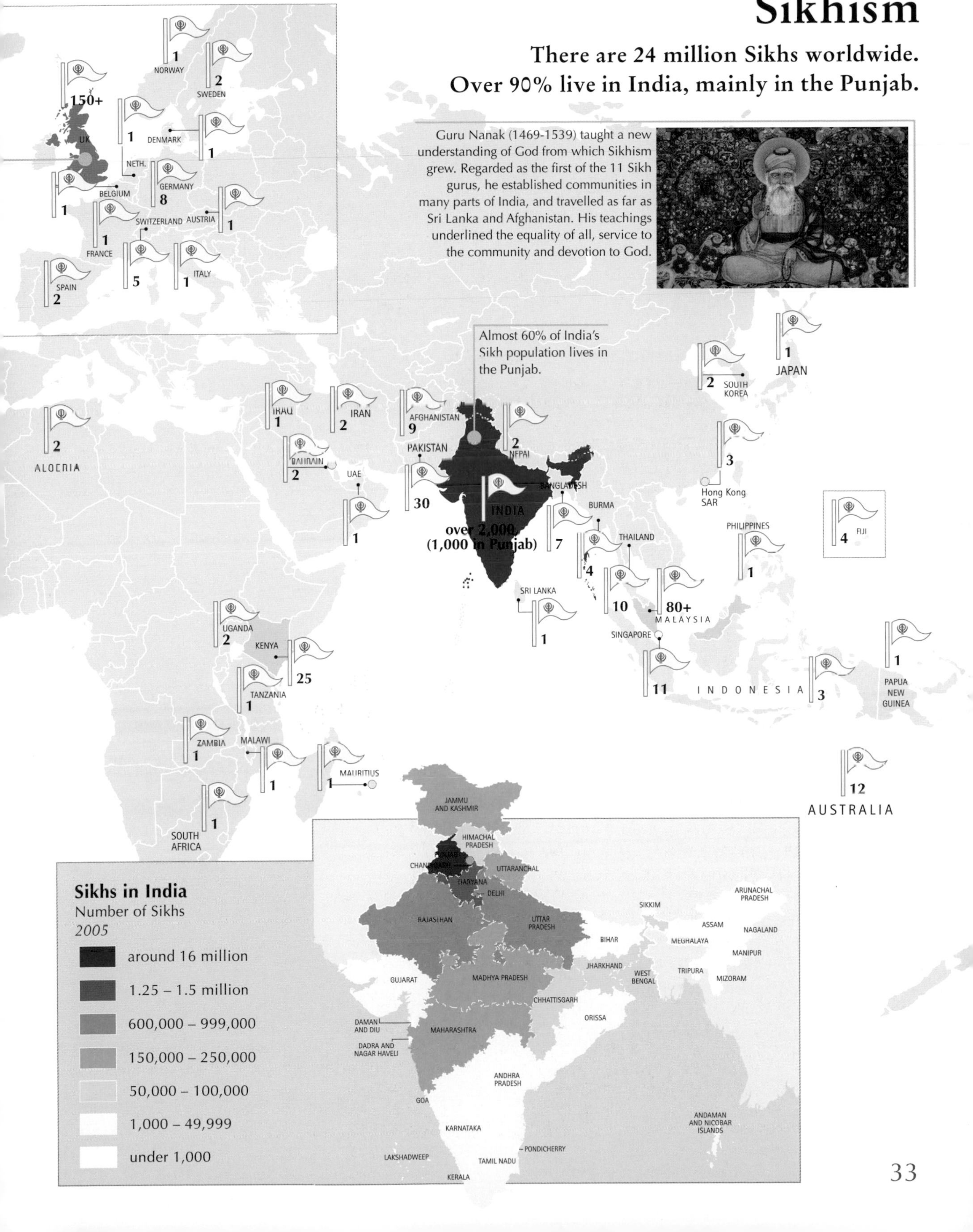

Almost 60% of India's Sikh population lives in the Punjab.

UK 150+
NORWAY 1
SWEDEN 2
DENMARK 1
NETH. 1
BELGIUM 1
GERMANY 8
AUSTRIA 1
FRANCE 1
SWITZERLAND 5
ITALY 1
SPAIN 2
ALGERIA 2
IRAQ 1
IRAN 2
BAHRAIN 2
UAE 1
AFGHANISTAN 9
PAKISTAN 30
INDIA over 2,000 (1,000 in Punjab)
NEPAL 2
BANGLADESH 7
BURMA 4
THAILAND 10
MALAYSIA 80+
SINGAPORE 11
INDONESIA 3
SRI LANKA 1
HONG KONG SAR 3
SOUTH KOREA 2
JAPAN 1
PHILIPPINES 1
FIJI 4
PAPUA NEW GUINEA 1
AUSTRALIA 12
UGANDA 2
KENYA 25
TANZANIA 1
ZAMBIA 1
MALAWI 1
MAURITIUS 1
SOUTH AFRICA 1

Sikhs in India

Number of Sikhs

2005

- around 16 million
- 1.25 – 1.5 million
- 600,000 – 999,000
- 150,000 – 250,000
- 50,000 – 100,000
- 1,000 – 49,999
- under 1,000

JAMMU AND KASHMIR
HIMACHAL PRADESH
PUNJAB
CHANDIGARH
UTTARANCHAL
HARYANA
DELHI
RAJASTHAN
UTTAR PRADESH
SIKKIM
ARUNACHAL PRADESH
ASSAM
NAGALAND
MEGHALAYA
MANIPUR
BIHAR
JHARKHAND
WEST BENGAL
TRIPURA
MIZORAM
GUJARAT
MADHYA PRADESH
CHHATTISGARH
ORISSA
DAMAN AND DIU
DADRA AND NAGAR HAVELI
MAHARASHTRA
ANDHRA PRADESH
GOA
KARNATAKA
ANDAMAN AND NICOBAR ISLANDS
PONDICHERRY
LAKSHADWEEP
TAMIL NADU
KERALA

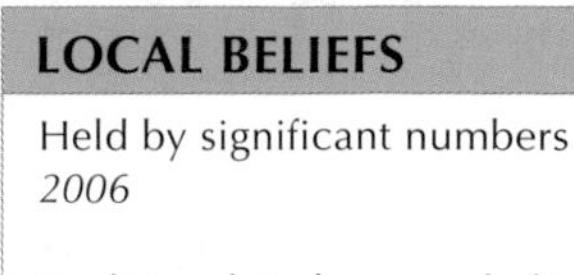

LOCAL BELIEFS

Held by significant numbers
2006

Traditional, indigenous beliefs:

widely diffused exist in pockets

Chinese indigenous religion:

 widely diffused

Shintoism:

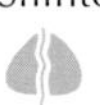 widely diffused

An inukshuk – a distinctive stone cairn built by the Inuit to mark high points and good hunting routes.

The terms 'traditional' and 'indigenous' distinguish those cultures and belief systems that are not part of a major world religion. While they often share key features, such as reverence for nature and veneration of ancestors, they do not adhere to any central tenets. Many belief systems – such as Australian aboriginal traditions – have been part of the same geographic setting for thousands of years. Some have travelled and been shaped by what is loosely called Shamanism, thought to have originated in Siberia and to have migrated with movement of peoples into the Americas.

The indigenous religion of China, which fuses Daoism, Buddhism, folk religion and Confucianism, is the majority religious practice of the country. In Japan, Shinto rites are widely practised, and even though an individual may profess a different 'personal' religion, Shintoism is often the 'family' religion.

Traditional religion is still widely practised throughout Sub-Saharan Africa. In the Caribbean and Latin America there are more than 3 million adherents of indigenous religions, from the 14,000 strong Huichol community in northern Mexico to the native peoples of the Amazon basin.

Traditional Beliefs

There are more than 250 million adherents of traditional beliefs worldwide, excluding the indigenous religions of China and Japan. There are also many who belong to a major world religion while continuing to hold traditional local beliefs.

A Siberian Shaman in ritual clothing inviting the spirit of the fire to give a blessing.

A tayakh, erected at a special worshipping place in the mountains of the Mongolian Altai. Ribbons are tied to the pole and offerings placed at its base.

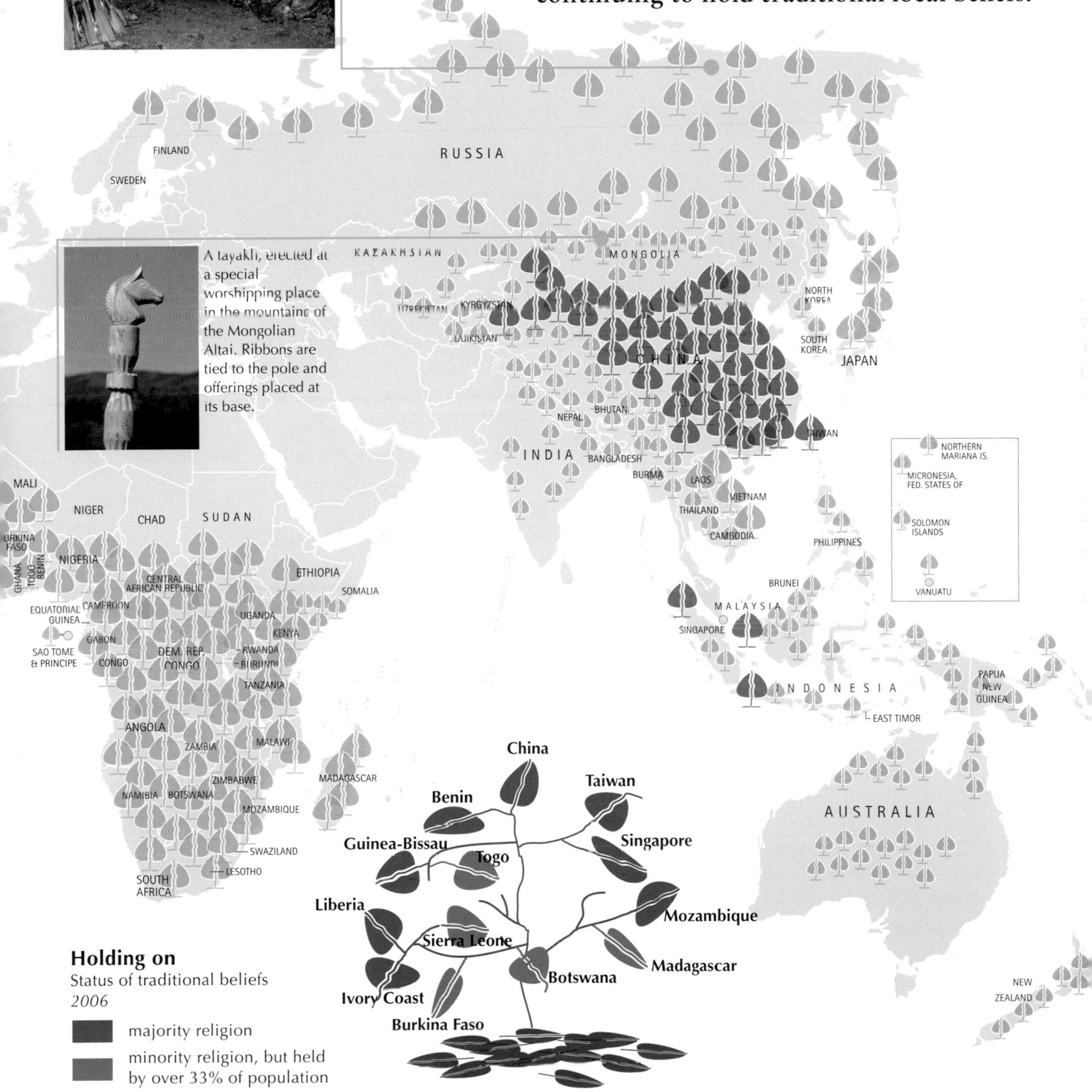

Holding on

Status of traditional beliefs

2006

- majority religion
- minority religion, but held by over 33% of population

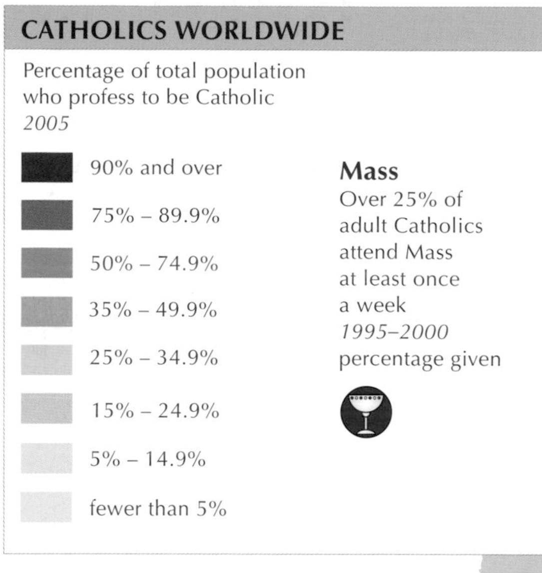

The Catholic Church is the largest Christian denomination in the world, with over 1 billion baptized members.

While the number of priests worldwide and in Latin America remains steady, in 2003 the Vatican recorded a decrease in European priests and an increase in priests from Africa and Asia. The number of nuns worldwide has dropped significantly, particularly in Europe and North America, where there were 15,000 fewer in 2003 than in the previous year. In contrast there was an increase of nearly 4,000 nuns in Asia and 1,285 in Africa. Overall, however, Europe and the Americas still have the largest number of nuns, with 74 percent of the world total of 776,269.

The Catholic Church is governed by the Holy See in Rome, officially known as the State of the Vatican City. It is the world's only sacerdotal government and as such has official diplomatic representation. There are three types of pontifical representative: nuncios, for nations with a Catholic majority; pro-nuncios, for nations with a Catholic minority; and, permanent observers or delegates who are appointed to certain international organizations.

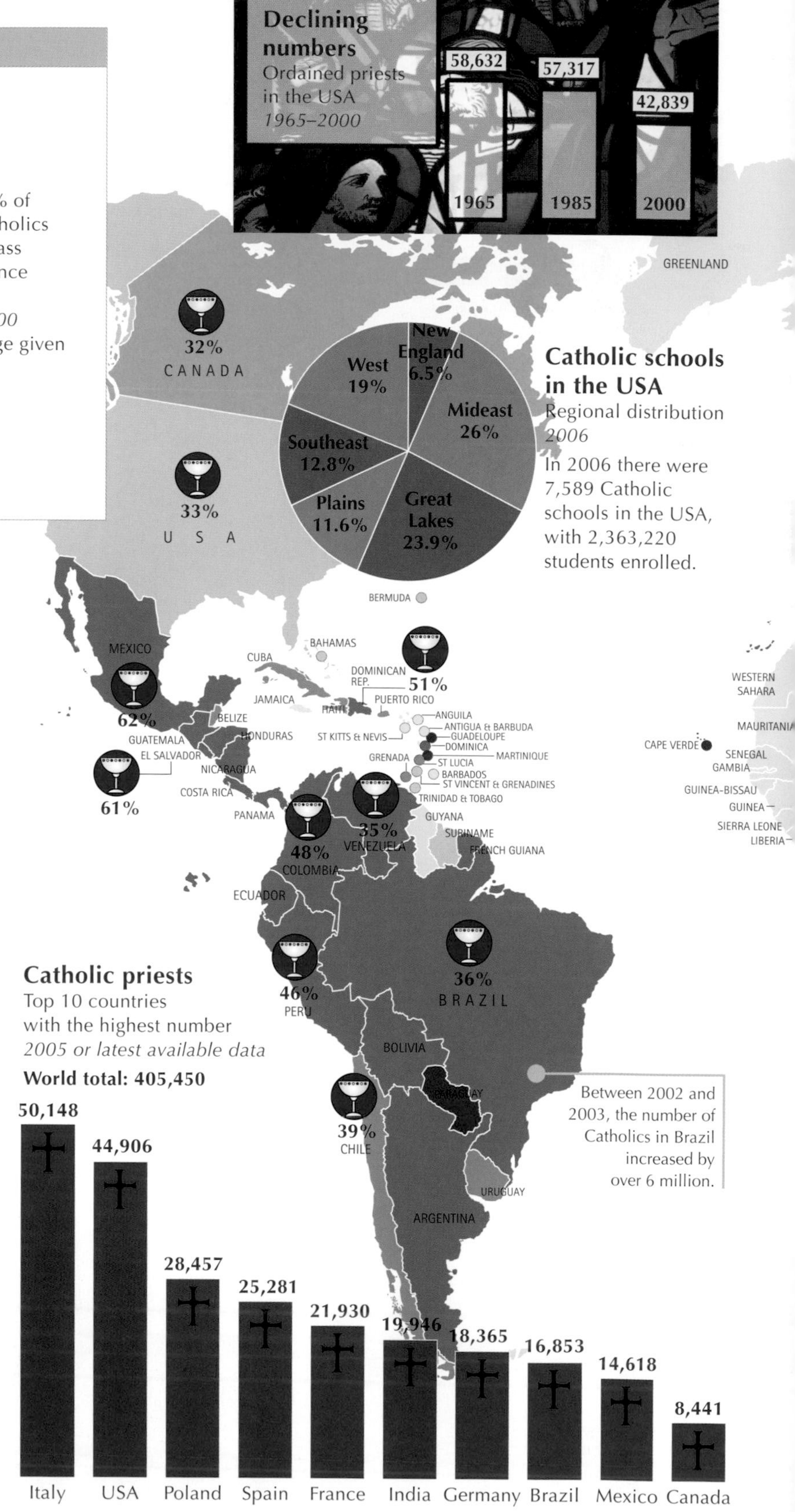

Catholicism

There are more than 1 billion Roman Catholics worldwide. 200,000 schools, serving more than 52 million students, operate under Catholic Church auspices.

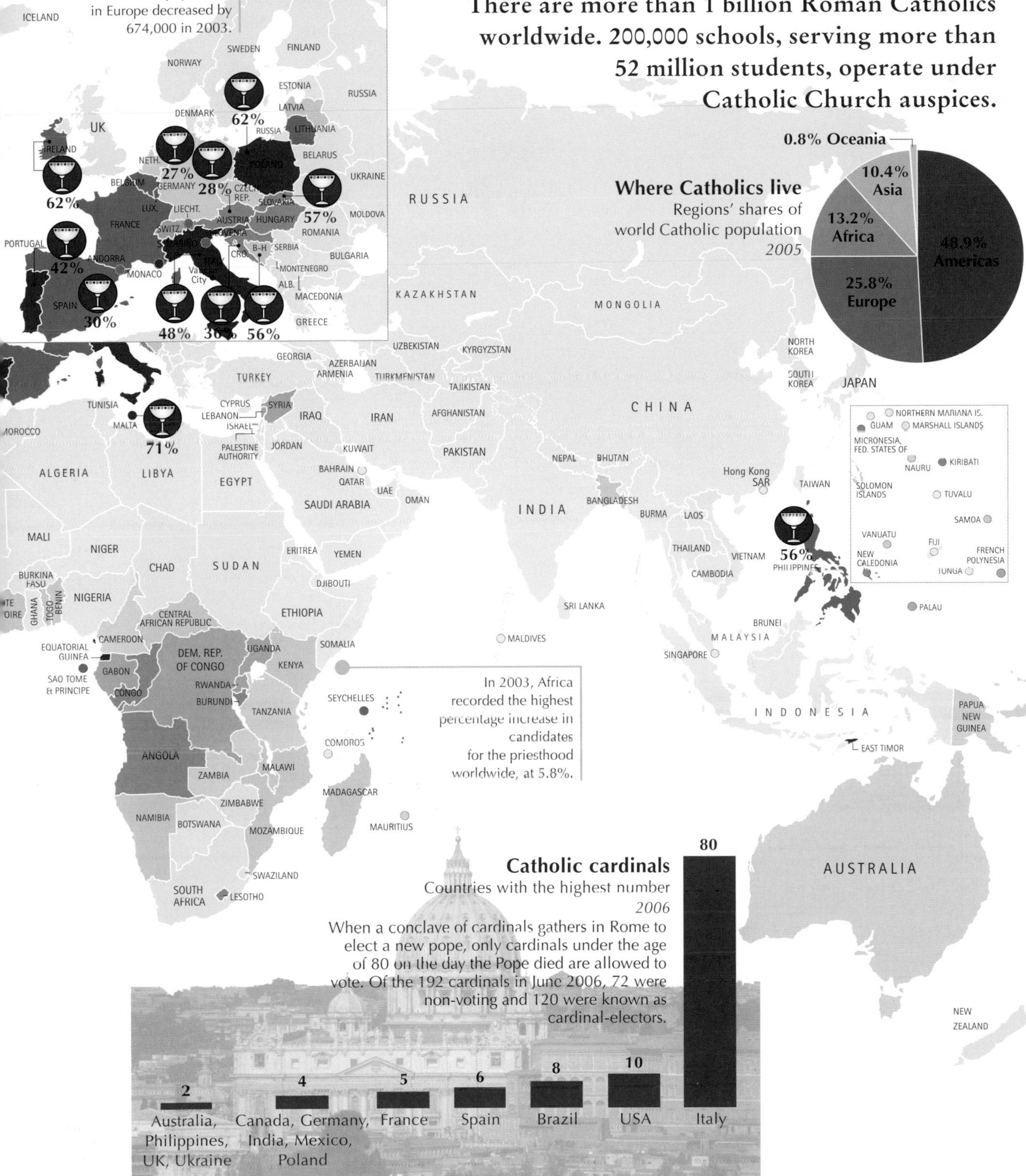

CENTRES AND HEARTLANDS

Selected recent movements
2006

- Church of Scientology
- World Plan Executive Council *Transcendental Meditation*
- Brahma Kumaris World Spiritual University
- The Family Federation for World Peace and Unification *Unification Church*
- School of Economic Science

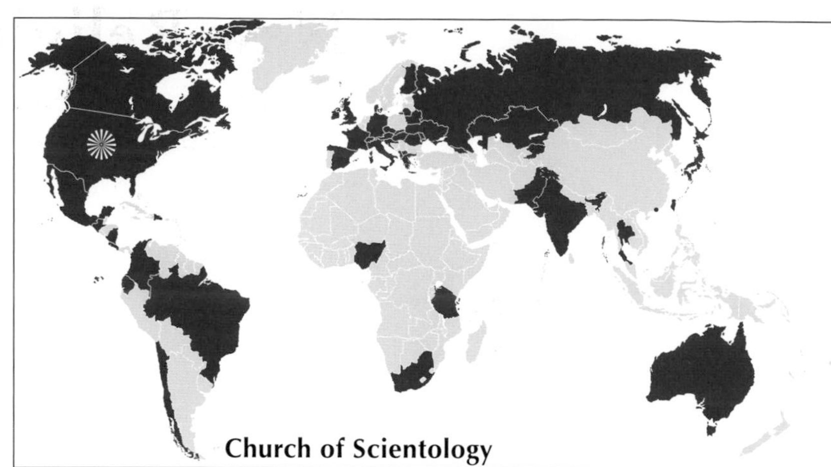

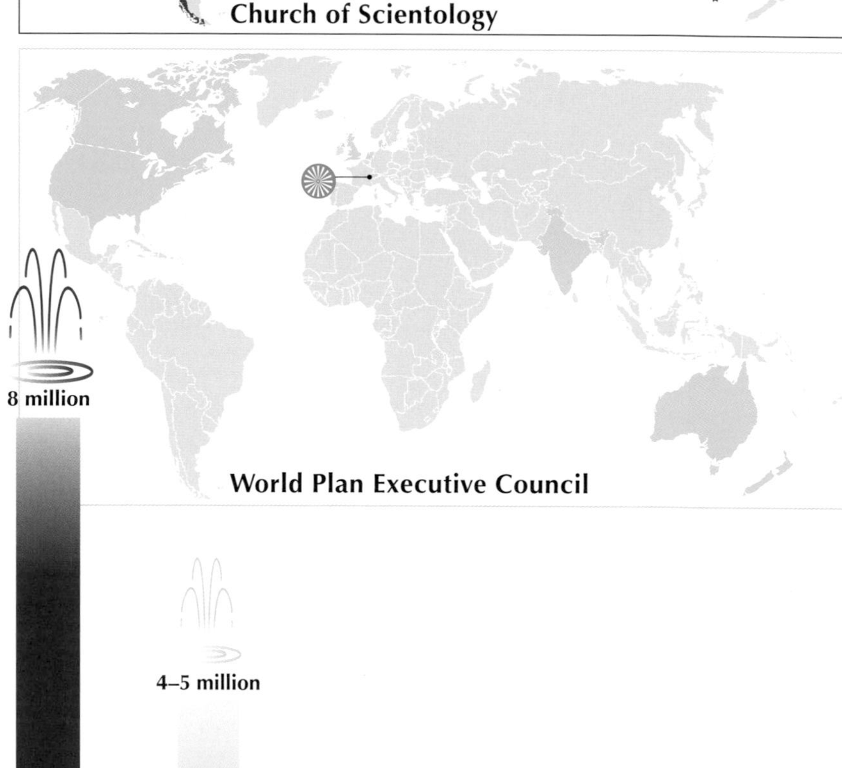

New religious movements are religious or spiritual groups not officially recognized as standard denominations or Churches, but their definition is one of the most controversial in the field of religious studies. Some claim to have traditional roots and others are entirely new. All have their strongest appeal in predominantly Christian cultures where secularism has created crises of identity.

The rise of new religious movements was aided by the growth of economy air travel from the 1960s, which made it possible for young people, swamis and teachers to travel extensively. Indonesia, with its long tradition of religious pluralism, has hundreds of new religious movements and is the only country in the world to offer them formal recognition and official protection.

Shown here is a selection of recent movements with no historical roots, such as the Church of Scientology, or movements not accepted by their supposed parent faith, such as the Church of Unification, which is not recognized by mainstream Christians. Not covered are the thousands of people who belong to local new religious movements, who sometimes exceed the number of those involved in international movements.

Selected 20th-century religious movements

Number of adherents and date founded

8 million

4–5 million

450,000

250,000

100,000

Church of Scientology 1954 **heartland USA**

World Plan Executive Council *Transcendental Meditation* 1958 **heartlands USA, India**

Brahma Kumaris World Spiritual University 1937 **heartland India**

Unification Church 1954 **heartland South Korea**

School of Economic Science 1938 **heartland UK**

New Religious Movements

The migration of people and ideas fosters new religious movements. Most, but not all, have their origin in a major world religion. Many are very localized, while a few are now international.

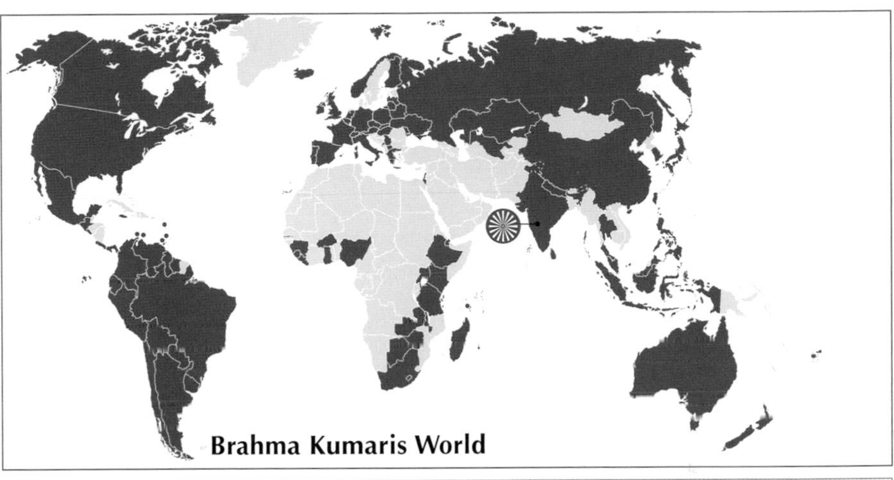

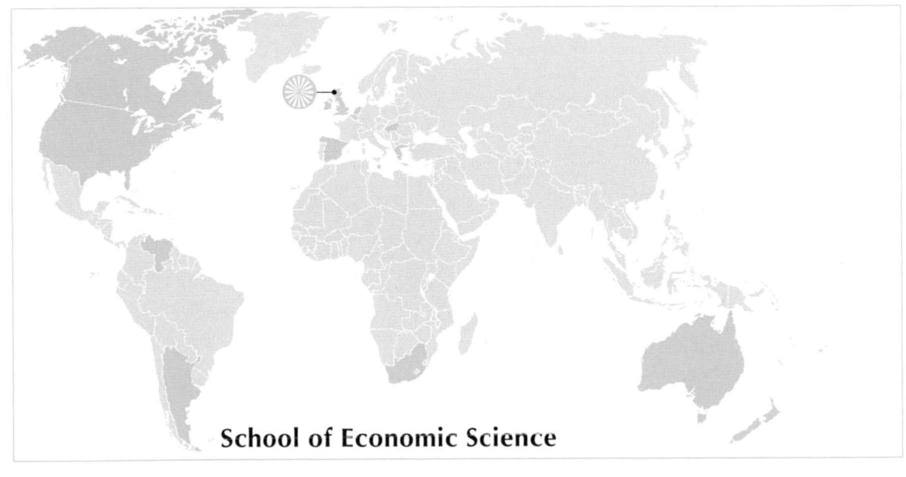

The creation of pan-Native American movements in the USA and Canada to formulate a common Native Religion is a conscious attempt to return to roots and reclaim identity. This has flowed through American society and into Europe, particularly in the New Age milieu which has adopted elements of traditional practice such as shaman groups, medicine lodges and sweat lodges.

Over the past few decades, specific forms of Paganism have been arising in different countries around the world as people have sought to retrieve what they perceive to be a more 'authentic' religious tradition that is closely tied with the culture or mythology of their nation. Celtic Druidry draws inspiration from the Celtic traditions of England, Wales, Scotland, Ireland and Brittany. Heathenry, a branch of Paganism drawing on Anglo-Saxon and Norse mythology, seeks to revive Northern Europe's pre-Christian traditions.

Shamanism, with its underlying belief that an individual in an alternative state of consciousness can travel between various different worlds, is active in Siberia, in the USA amongst Native Americans, and in Latin America, Indonesia, Asia and Europe. Practised by indigenous people worldwide, Shamanism is also emerging as a new spiritual movement in New Age and Pagan communities in many industrialized countries.

The slave trade between Africa and the Americas, from the 16th to the 19th centuries, blended African indigenous religions with Christianity and the traditions of the Caribbean, Central and South America. In Brazil, syncretized religions are followed by almost 5 percent of the population, and 15 percent are engaged in rites while also belonging to a major faith. Many of the followers of Santeria in Cuba are also professing Catholics and, in Haiti, Vodoun is officially recognized as a religion.

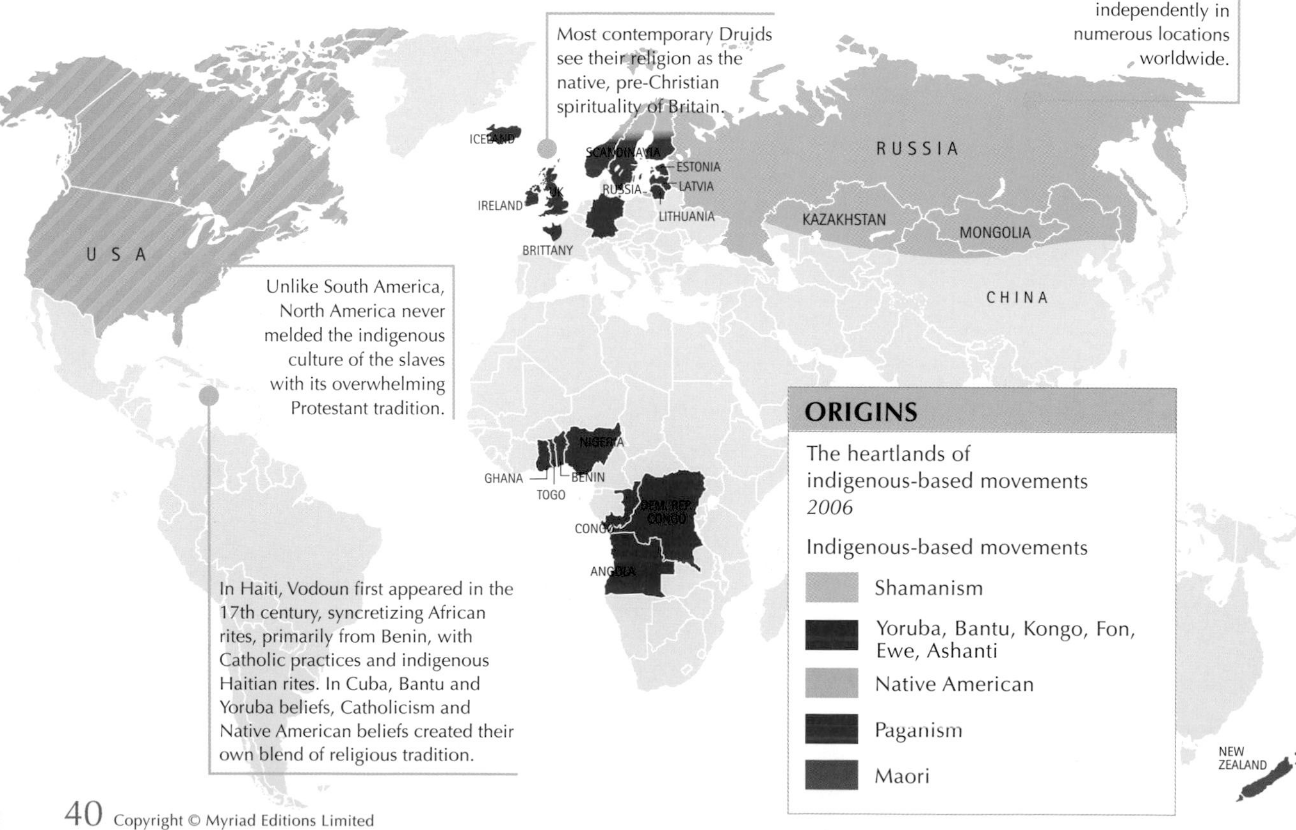

New Departures

As indigenous religions have spread, they have blended with traditions encountered en route or at their destination. Other traditions have been revived to create new identities or strengthen old ones.

NEW DEPARTURES

THE GLOBAL ROUTES OF INDIGENOUS-BASED MOVEMENTS 2006

MOVEMENT	GROUP	DEPARTING FROM	VIA	DESTINATION
SHAMANISM/NATIVE AMERICAN RELIGIONS	NATIVE AMERICANS	USA/CANADA		NORTHERN EUROPE, WESTERN EUROPE
CORE SHAMANISM	AMERINDIANS	SOUTH AMERICA		USA, EUROPE, AUSTRALIA, NEW ZEALAND
MAORI TRADITIONS	MAORI	NEW ZEALAND		USA, WESTERN EUROPE, AUSTRALIA
VODOUN	YORUBA, FON, EWE, KONGO, ASHANTI	NIGERIA, BENIN, CONGO, TOGO, GHANA, ANGOLA	HAITI	WEST INDIES, USA
SANTERIA	YORUBA/ BANTU	NIGERIA, BENIN, CONGO, TOGO	CUBA	CARIBBEAN, USA, ARGENTINA, BRAZIL, COLOMBIA, MEXICO, PUERTO RICO, VENEZUELA, NORTHERN EUROPE
CANDOMBLE	YORUBA, BANTU, KONGO, FON, EWE, ASHANTI	NIGERIA, BENIN, CONGO, TOGO, GHANA, ANGOLA	BAHIA, NORTH-EAST BRAZIL	BRAZIL
PALO MAYOMBE	KONGO	CONGO, ANGOLA	BRAZIL	CUBA, MEXICO, NORTHERN SHORE OF SOUTH AMERICA
SPIRITISM	KARDECISM	FRANCE	BRAZIL	PUERTO RICO, USA
UMBANDA	AFRO-BRAZILIAN/ AMERINDIAN/ KARDECISM/ CATHOLICISM	RIO DE JANEIRO (FOUNDED 20TH CENTURY)		BRAZIL
PAGANISM	CELTIC DRUIDRY	UK		NORTHERN EUROPE, USA, AUSTRALIA, NEW ZEALAND
PAGANISM	HEATHENRY	GERMANY, ICELAND, SCANDINAVIA		UK, USA, AUSTRALIA, NEW ZEALAND

NON-RELIGIOUS AND ATHEISTS

People professing no religion as a percentage of total population
2005

- 20.0% or more
- 15.0% – 19.9%
- 10.0% – 14.9%
- 5.0% – 9.9%
- fewer than 5.0%

Atheists
as percentage of population

10% or more

5% – 9%

Humanists

humanist or secular organizations based in country

The reasons why people profess to have no religious affiliation can be diverse, and include lack of interest, a commitment to free-thinking, agnosticism (being undecided), or simply not wishing to be labelled with any mainstream religion. Most tend not to be anti-religious; they are just not pro-religious.

The number of professed atheists – those who deny the existence of God – is very small, except in a few countries, primarily the former or currently communist countries, where decades of atheist education has made an impact.

Countries where humanist or ethical societies exist often have a strong tradition of opposing the role and power of religion within their culture, although they are not necessarily against religion as such. This is especially a feature of Western countries, due in part to the Enlightenment movement in the 18th century, and its struggle with religious hegemony. There is great interest in such organizations in India, arising from the drive to maintain a secular culture within which all religions can play a part, with no single religion dominating.

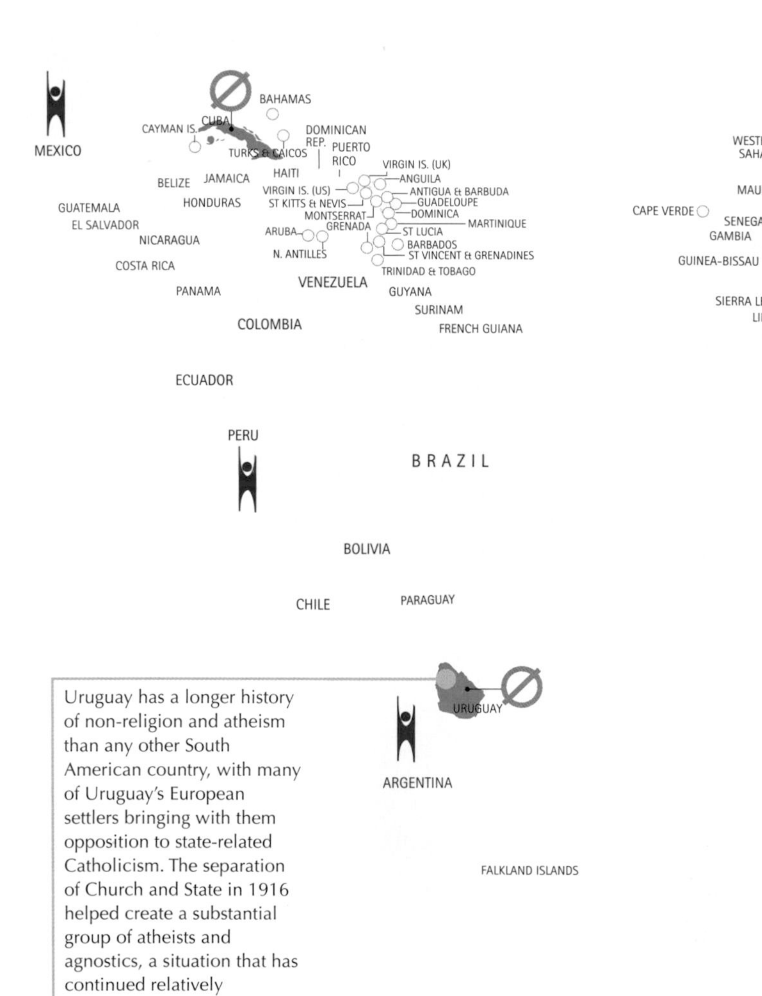

Uruguay has a longer history of non-religion and atheism than any other South American country, with many of Uruguay's European settlers bringing with them opposition to state-related Catholicism. The separation of Church and State in 1916 helped create a substantial group of atheists and agnostics, a situation that has continued relatively unchanged.

Non-Believers

Over 10% of people claim no allegiance to a religion. Many of these are undecided, but some are atheists, who deny the existence of God.

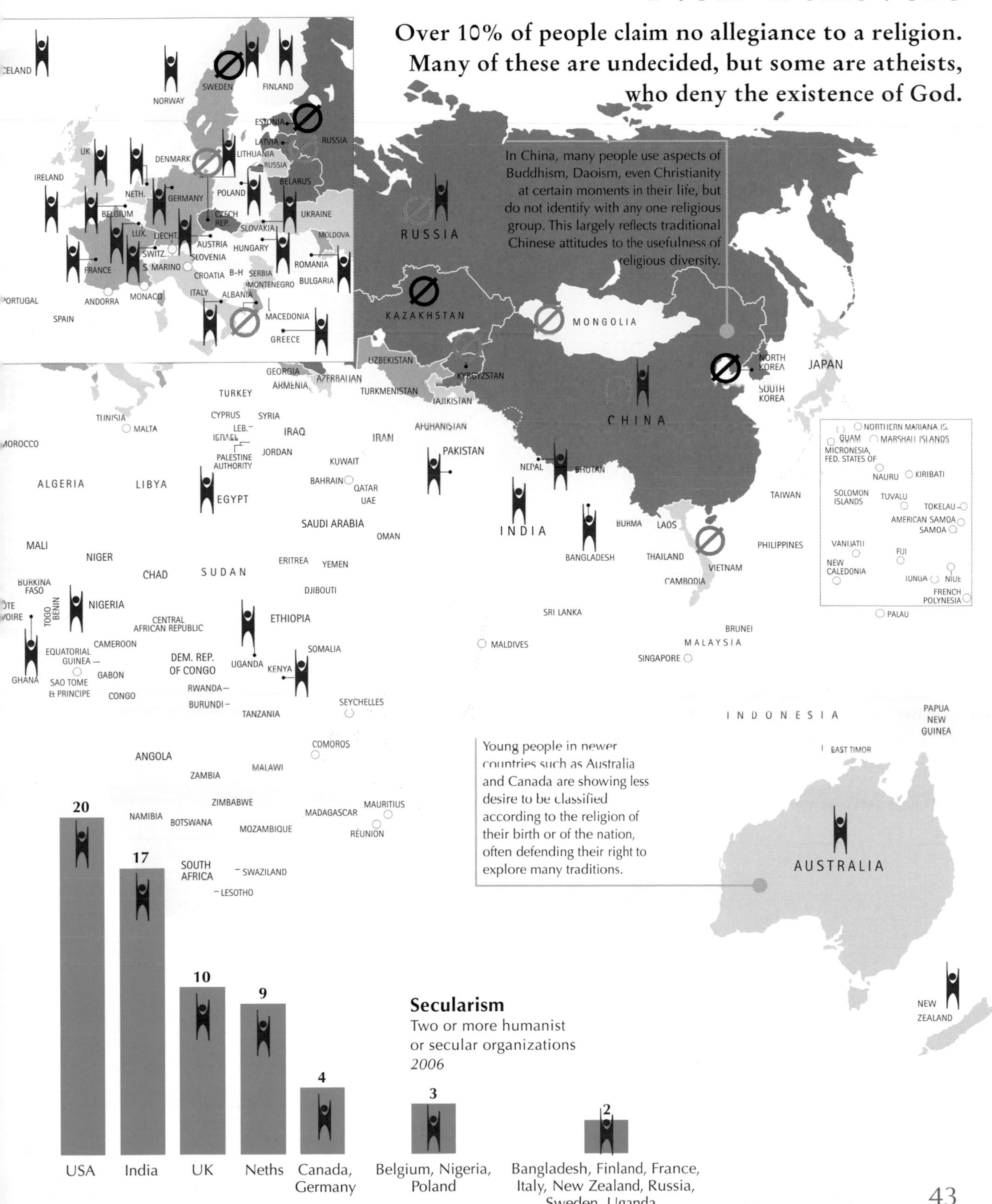

Part Three STRUCTURES

All successful religions survive through the structures they have created over time. The original founder may provide the energy and vision, but it takes the next few generations to ensure the survival of the teachings. The Buddha, for example, was one teacher among hundreds, but his message survived because he created what are known as the Three Jewels: the Buddha, the Dharma (teachings and the true path) and the Sangha (the community of monks and nuns to preserve and pass on the teachings).

Strong structures are especially important in the case of the major missionary faiths: the financing of Christian activities by the faithful, and the structure of Islamic Law are two examples. The apparatus involved in spreading the message plays a major role. Faiths have for a long time funded the publication of core texts, but increasing emphasis is being placed on media such as TV, radio and the internet as a means of structuring people's encounter with core teachings.

The relationship between State and religion can be controversial. For many people, the idea that a state should support a specific religion – through either funds or state-provided single-faith religious education – is anathema. The 18th-century insistence on the separation of Church and State, first espoused by the fledgling USA and taken up by France in its revolution, has now become a core tenet of secularism. Yet in Islamic countries such a notion is itself anathema, and almost all are Islamic in law, practice and politics. In Europe, the old Protestant ideal that the religion of the people should be reflected in a State Church and with a religious monarch has begun to diminish, but still is the norm for many countries in Northern Europe, while the Catholic countries of the South of Europe hold on to their position as officially Catholic.

Elsewhere, most states that have emerged over the last 100 years, including Australia, Canada, Turkey, India and Angola, have opted for secular constitutions, in which all faiths are recognized and protected, but none is chosen as The Faith.

It is through religious education that one can sometimes catch a glimpse of this complex relationship. For example, of the countries that emerged from the collapse of the Soviet Union and its satellites, most have opted for the secular approach. Kyrgyzstan, however, will not permit the teaching of religion or atheism in its state schools, while Afghanistan has opted for Islamic education.

Muslim female pilgrims visiting the citadel of Bam, Iran, prior to the earthquake that destroyed much of this ancient citadel in December 2003

FREEDOM AND RESTRICTION

State attitude to religion
2006 or latest available information

- discriminates against all religions and interferes with religious freedom
- favours religion of majority and interferes with or limits freedom of other religions
- favours religion of majority but tolerates other religions
- tolerates all religions
- state declared atheist in law
- state religion established in law
- state recognizes more than one religion or religious group
- monarch must be of given religion
- head of state or government must be of given religion

The separation between Church and State only emerged in the late 18th century, as a result of the American Revolution. Many people had fled or migrated to America to escape coercion into a religion that was not theirs, and the idea of a State Church was anathema. Since then, a state independent of, and unaffiliated to, any particular religion has become the norm. Countries that emerged from colonialism during the 20th century have largely opted to be secular states.

Within the Muslim world, however, such a divide between religion and State is almost unknown. Islam is a way of life that encompasses legal systems and administration, as well as prayer, instruction and morality. Islam is also recovering ground previously lost to secularism. For example, Iran and Afghanistan have both turned their backs on secular philosophies and reverted to Islamic states.

Europe is in a state of transition. Northern Europe tends to be Lutheran or Anglican – traditions that helped create the notion of a distinct nation during the Reformation of the 16th century – but in recent years the Church–State partnership has weakened, exemplified by Sweden, which disestablished its Church in 2000. Southern Europe is Catholic; the Catholic Church held on to these areas in the face of the rise of the Protestant threat. It is the Holy See that negotiates the Church–State relationships in these countries, as with all majority Catholic countries, leading to a very different relationship from that between Protestants and the State.

Elsewhere in the world, religious ties are once again being developed or proposed. In India, one of the major parties overtly campaigns for Hinduism to be the state religion. In Burma and Sri Lanka, Buddhism is used as a rallying point for nationalism in the struggle between the majority culture and minorities.

State Attitudes to Religion

Nearly a quarter of the world's states have formal links with a religion. Some have links with more than one. A few actively discriminate against all religions.

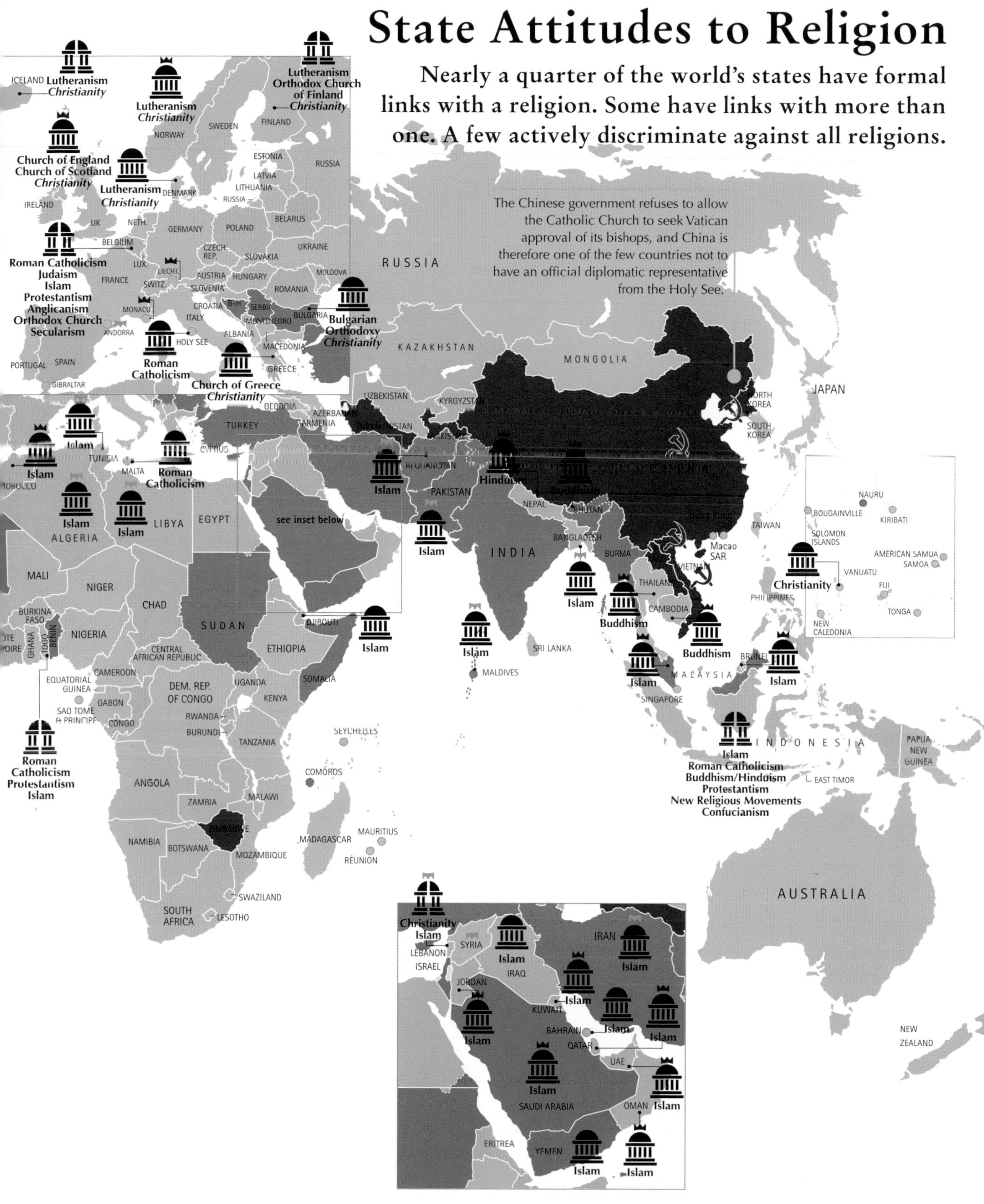

CHURCH INCOME

Annual income received by Christian Churches from members
2005
US$ million

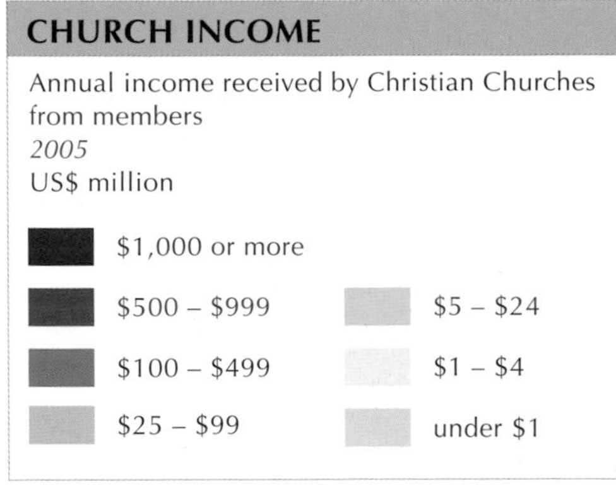

Christians make up a third of the world's population, but receive more than half the world's total annual income. This is because they are concentrated in the industrialized world, and their average annual income, at around $8,000, is well above the world average. This figure disguises an unequal distribution, however. While half of all Christians live in affluence, and over a third are comparatively well off, 13 percent live in poverty.

Affiliated church members give over $270 billion to Christian causes. While 40 percent goes towards the running of their denominations and their local church, 60 percent is for agencies founded by Christian groups and run by Christians to support welfare programmes, health and education facilities, religious programmes, aid and development projects, and other charities or foundations. The finances of these parachurch agencies are independent of the Churches.

Christians are also heavily involved in financial support of social and development programmes beyond or outside the boundaries of their Church, giving $27.1 billion to secular charities that provide, for example, famine relief, hospitals and medical research.

While North Americans and Europeans make the largest individual financial contribution to their Churches, Christians in Africa, Asia and Latin America make a substantial contribution of their time and skills.

Christian Finance

The world's 2.1 billion Christians donate more than $297 billion a year to support Christian and non-Christian causes.

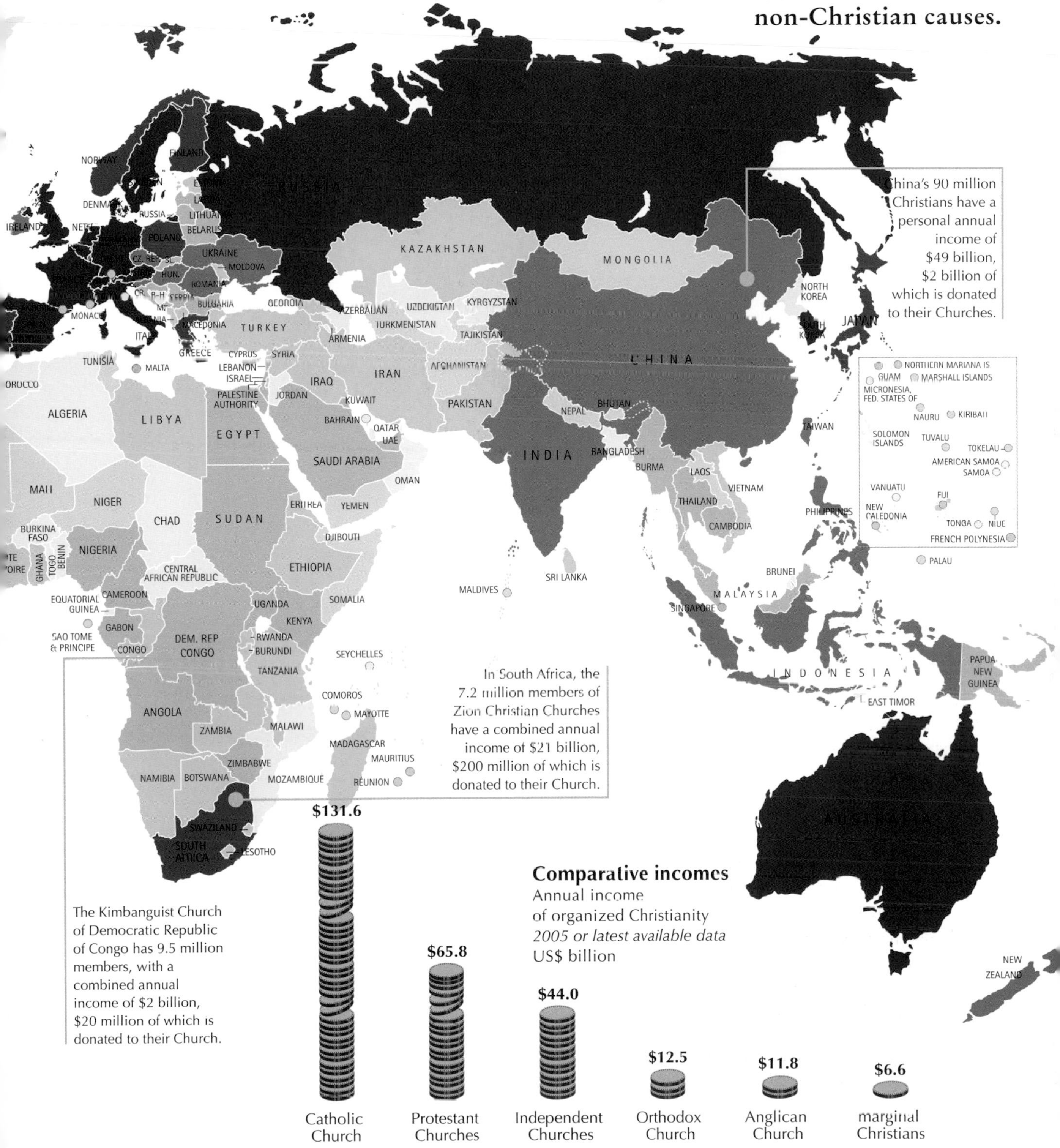

China's 90 million Christians have a personal annual income of $49 billion, $2 billion of which is donated to their Churches.

In South Africa, the 7.2 million members of Zion Christian Churches have a combined annual income of $21 billion, $200 million of which is donated to their Church.

The Kimbanguist Church of Democratic Republic of Congo has 9.5 million members, with a combined annual income of $2 billion, $20 million of which is donated to their Church.

Comparative incomes
Annual income of organized Christianity
2005 or latest available data
US$ billion

$131.6 Catholic Church
$65.8 Protestant Churches
$44.0 Independent Churches
$12.5 Orthodox Church
$11.8 Anglican Church
$6.6 marginal Christians

STATE PROVISION OF RELIGIOUS EDUCATION

2006

Type of religious education offered

- single faith
- multi-faith with emphasis on majority religion
- multi-faith
- no state religious education offered
- no data
- religious education not compulsory
- religious education under debate

Until the 1960s, attitudes towards religious education were largely polarized. In many countries it was not offered at all, either because of the separation of Church and State (as in France and the USA), or for ideological reasons (as in China and the USSR). In others, it was used to propagate the majority religion of the country (as in Saudi Arabia and the UK).

Over the last 50 years there has been a shift and, where religious education exists, multi-faith education is increasingly the norm, although in many countries there is still a lively debate about whether the majority religion should be accorded special status.

The former communist countries of Europe have undergone a huge shift since the early 1990s. From a complete ban on religious education under communism, a number of them have moved through a period where responsibility for religious education was held by the dominant religious tradition – Russian Orthodoxy in Russia, and Catholicism in Poland – to a situation where there is now a demand for multi-faith education. This has led to struggles between the Churches and the state education authorities.

In many countries where there is no state religious education, religious groups run classes within the context of faith schools.

Religious Education

The relationship between State and religion is often revealed by a country's attitude to religious education.

Council of Europe

Members

2006

The Commission of Foreign Ministers of the Council of Europe recommends principles for dealing with religion in public education in all 46 member states. This essentially calls for all 46 countries to have multi-faith state religious education in publicly funded schools. Formal inter-governmental recognition, on this scale, of the role of religious education in creating a multi-cultural and religious world has never previously occurred.

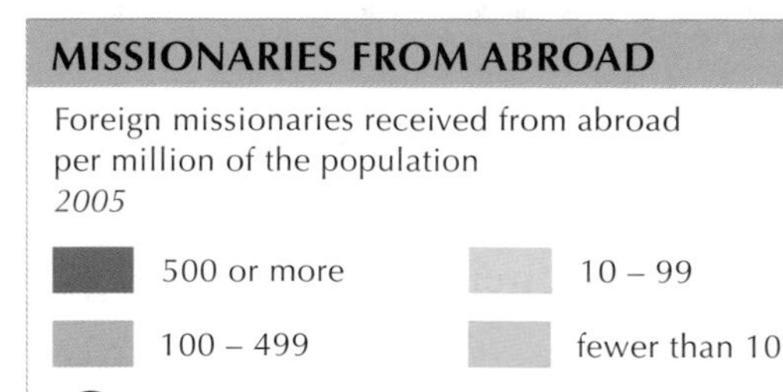

There are 1.6 million Christian missionaries, 99 percent of them working among existing Christians, and 70 percent in their home country. The 419,500 Christians working as foreign missionaries are managed by 4,100 mission boards and agencies. The annual cost of this work is $15 billion, provided largely by Church members. Just over 5 percent of Christian giving is for foreign mission work.

As well as evangelization and Christian renewal, missionary work includes education, the provision of health programmes, partnership in development programmes including agricultural and environment projects, and work with communities on justice and peace issues.

The largest numbers of missionaries still come from the traditional mission-sending countries in Europe and the Americas but, increasingly, former 'mission field' countries, such as South Africa, Nigeria and the Philippines, are sending missionaries to work abroad – sometimes back to the old, mission-sending countries themselves.

In countries where foreign missionary activity is restricted or prohibited for religious or political reasons, internal missionary activity may not necessarily be banned. Where there is state opposition or community hostility towards Christianity and the sending and receiving of missionaries, in practice small numbers of missionaries may be sent or received, usually serving as chaplains or in secular occupations.

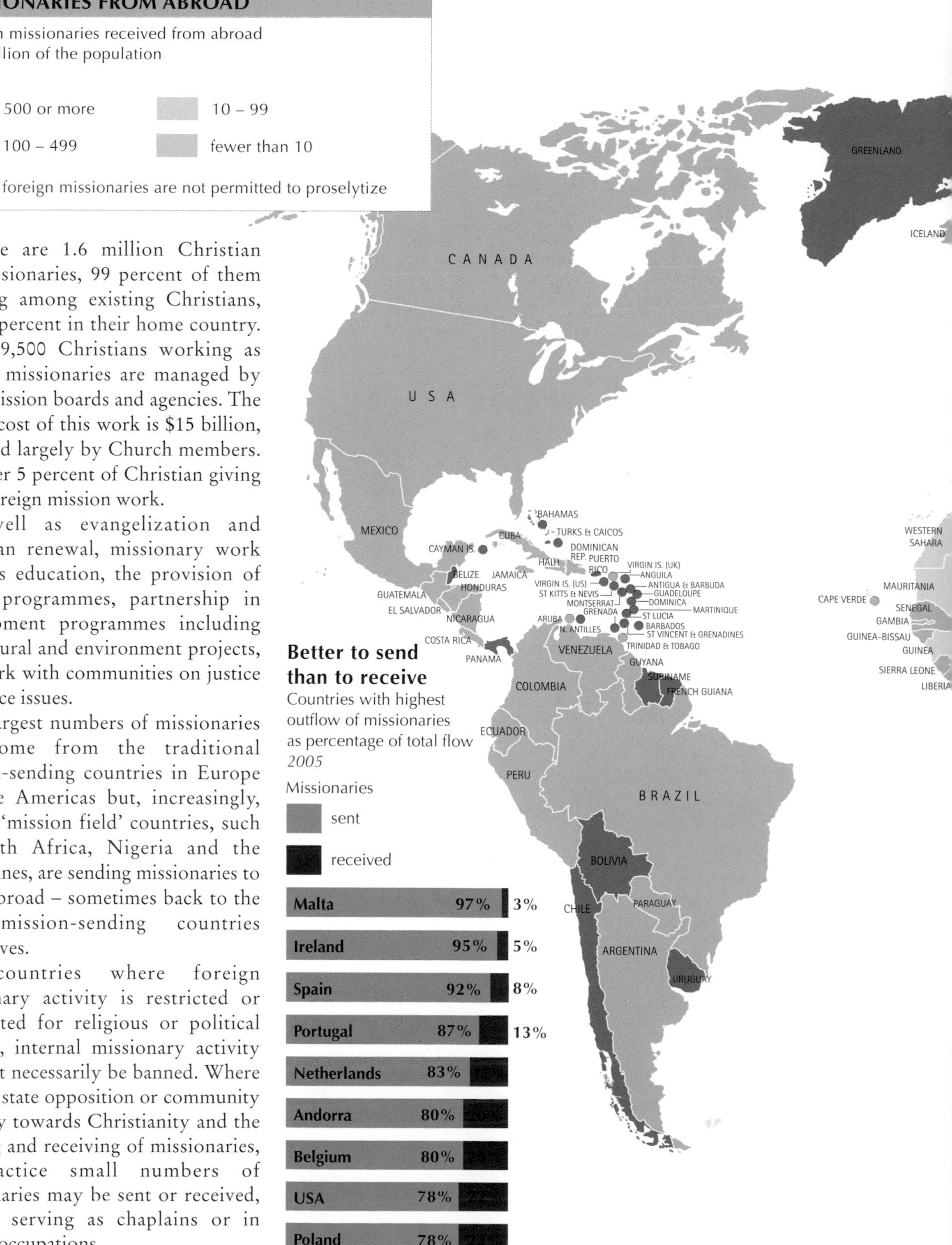

Christian Missionaries

The majority of Christian missionary work is among existing Christians. Most missionaries work in their country of origin.

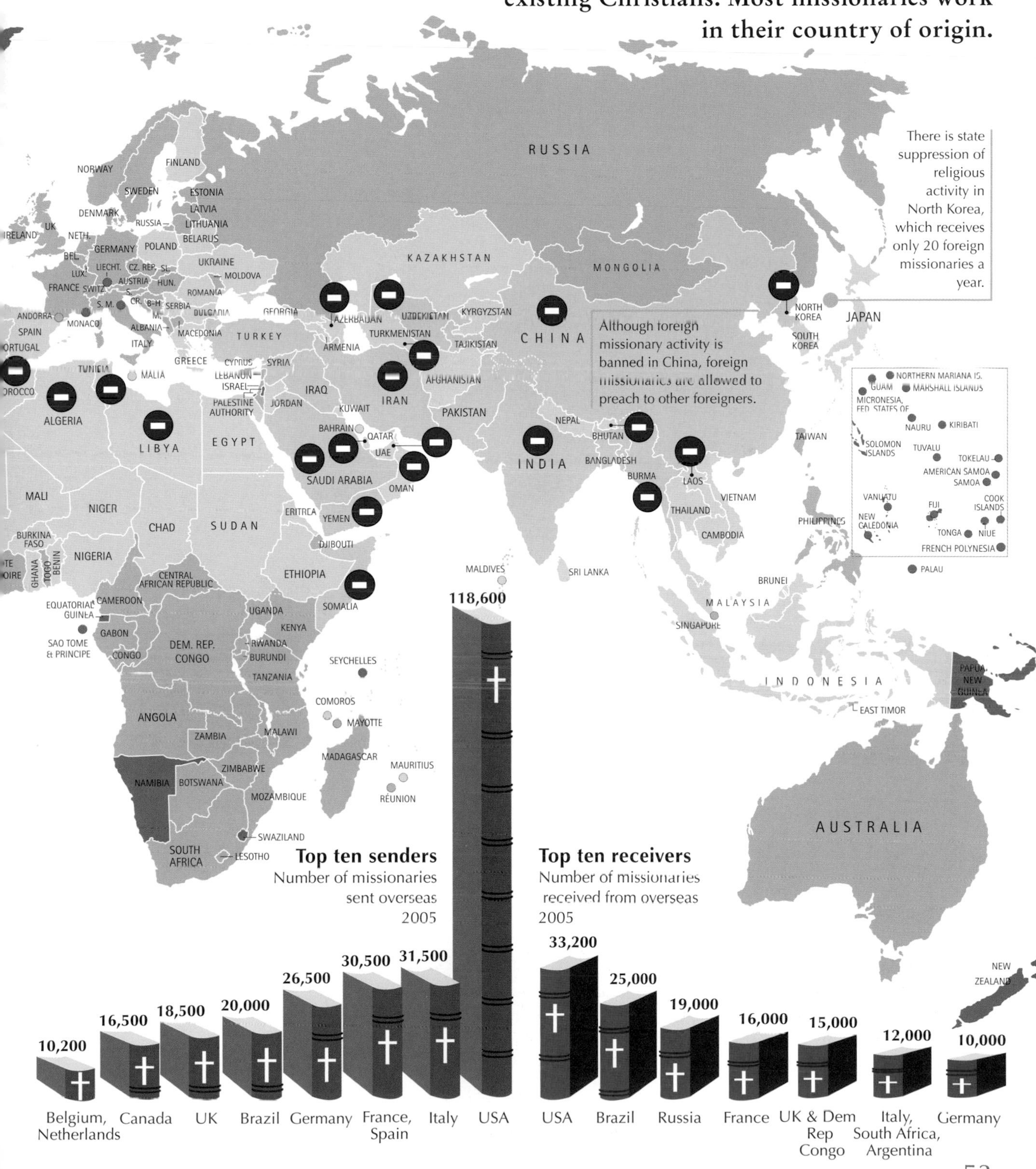

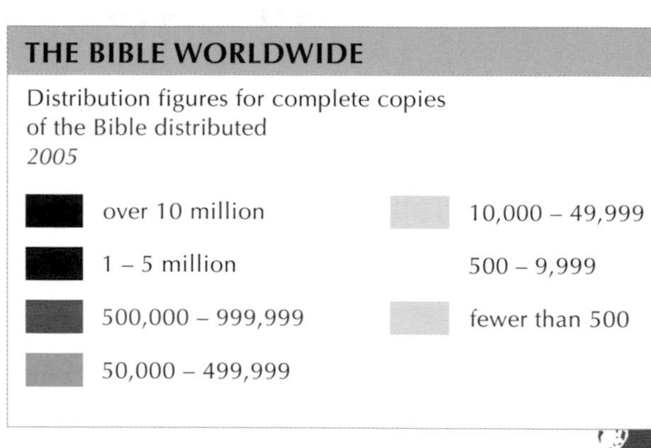

While 72 million bibles were distributed in 2005, another 1.5 billion were estimated already to be in place – assuming that a bible lasts for 20 years.

Agencies specializing in scripture dissemination keep account of six basic categories of scriptures: complete bibles; copies of the New Testament; portions of scriptures, which are usually a copy of one of the gospels; illustrated leaflets of up to eight pages; audio gospels; video gospels. These scriptures are distributed through commercial sales in bookshops, subsidized distribution in bible societies, churches and agencies, and free distribution by Gideons and similar organizations. In addition to publications of the Bible, some 45,000 Christian periodicals in 3,000 languages, with a combined circulation of 50 million, were produced in 2005. The USA publishes the highest number, with 8,000 titles.

Since the 1990s, the internet has become a major means of access to Christian scriptures, literature and discussion. The complete Bible first appeared on the internet in 1996. By 2005, it was available online in 110 languages. In 2005 there were 420 million personal computers in Christian use, and an active Christian internet network of 360 million people.

The distribution of Christian scriptures worldwide ranges from complete copies of the Bible to three-hour films and taped audio readings of scriptures called audio gospels. Global literacy among Christians is 87%.

In 2005, over 30% of all bibles were distributed in the USA.

A page from an illustrated Ethiopian Bible dating from the late 19th century.
In 350 CE sections of the Bible were translated into Ge'ez, an ancient Semitic language. The Bible Society of Ethiopia is trying to meet the demands of the country's 83 living languages in printed form and through their 'Faith comes by Hearing' audio programme.

The complete Bible has been translated into 426 languages, the New Testament into a further 1,115 languages, and a portion of the Bible into an additional 862 languages.

Nearly a quarter of all blind people have access to a Braille bible in their own language and audio recordings are available in more than 4,700 languages.

The Word

The Bible is the most printed and widely distributed book in the world. In 2005 there were over 1.5 billion bibles in circulation.

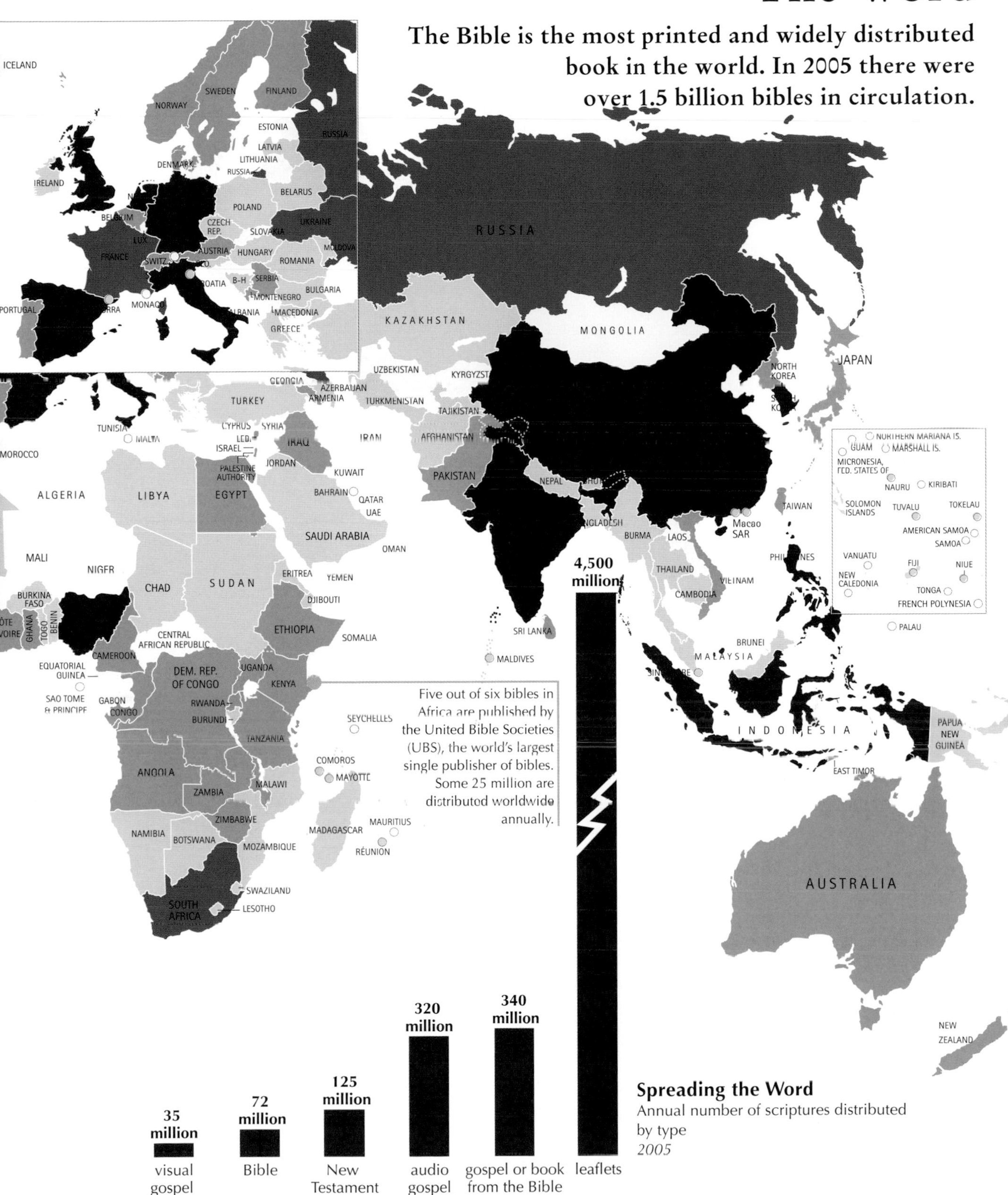

Spreading the Word
Annual number of scriptures distributed by type
2005

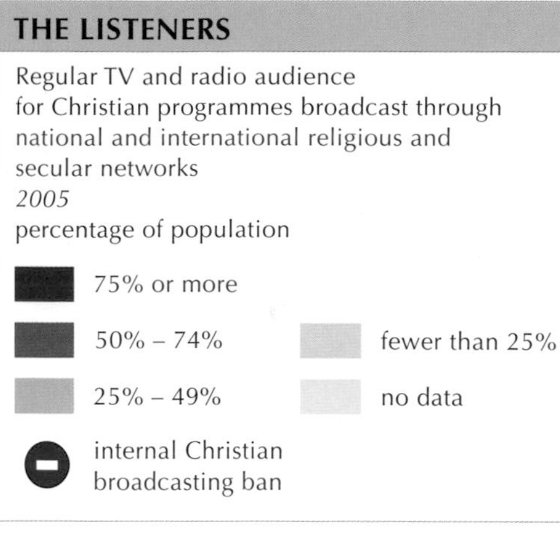

In 2005, broadcasts on a Christian theme reached a worldwide audience of almost 2 billion people via commercial, governmental or specifically Christian stations. Two-thirds of programmes were on radio and one-third on television. Christian agencies and networks spent $6 billion on Christian broadcasting.

The USA produces the greatest volume of Christian broadcasting and also has the highest regular audience. Worldwide, urban areas are likely to have more exposure to new forms of electronic media, but advances in satellite technology mean that programmes are increasingly being received in remote rural areas. Satellite transmitters are also reaching hitherto closed areas such as the Middle East, where certain states ban internal Christian broadcasting.

Since its beginnings in 1921, Christian broadcasting has been the most effective form of Christian evangelism. One of the most successful broadcast ministries has been the Lutheran Hour. Started in 1930, it is the world's longest-running Christian outreach radio programme. It is heard by 1.2 million people weekly over 800 radio stations across North America, but is also broadcast internationally on the American Forces Network, thereby expanding its audience to 40 million regular listeners.

Christian Broadcasting

Nearly 2 billion people listen to radio or television broadcasts on a Christian theme at least once a month.

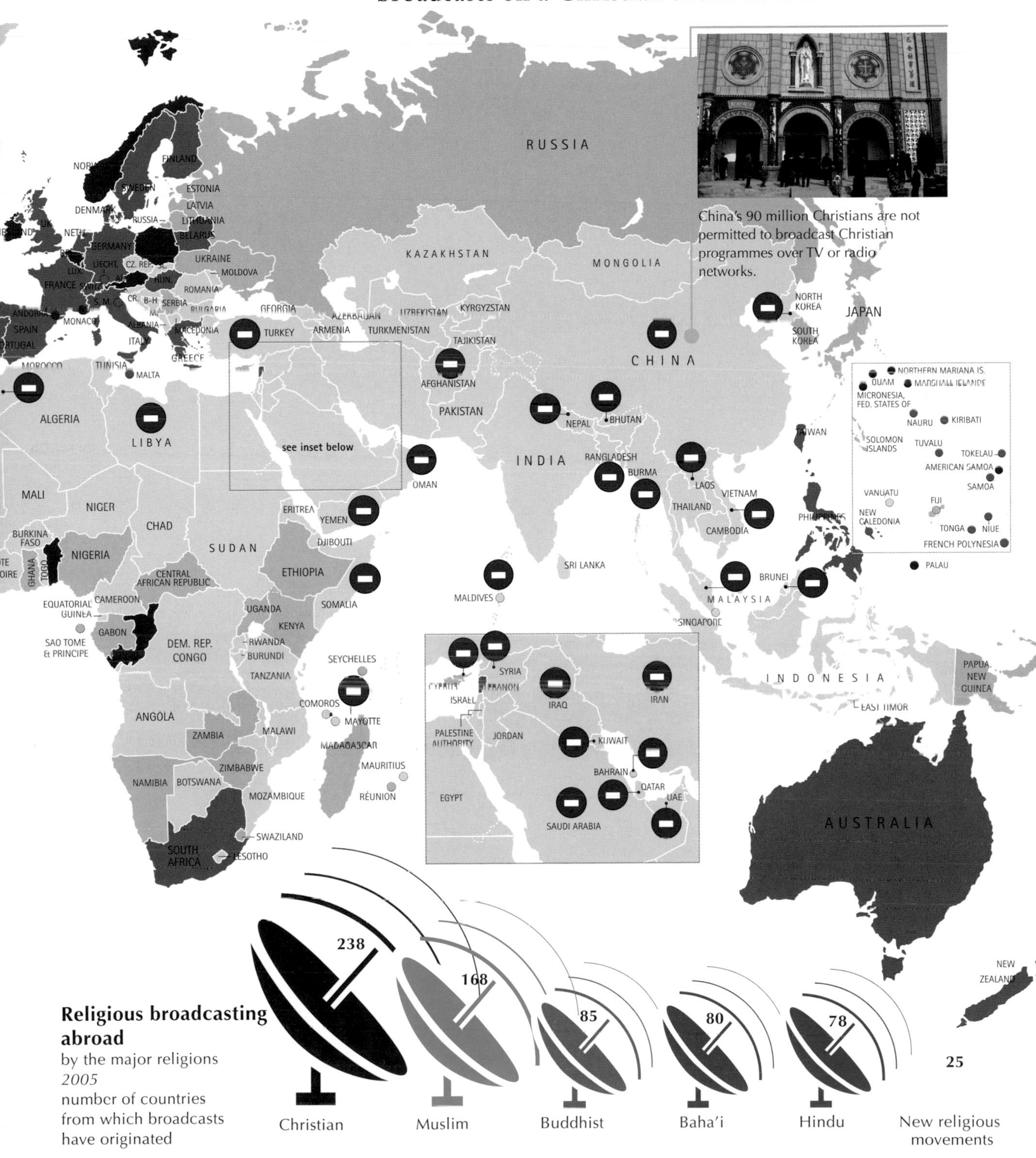

China's 90 million Christians are not permitted to broadcast Christian programmes over TV or radio networks.

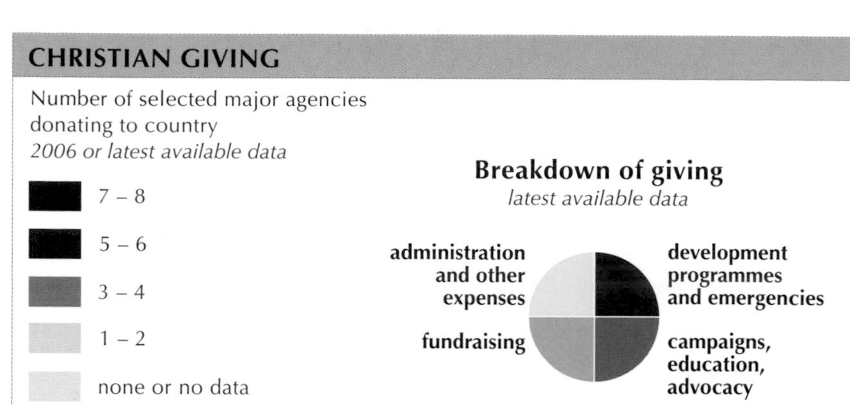

Historically, the major religions provided most, if not all, schools, clinics and social welfare structures worldwide. The rise of nation states and nationalism led to the State taking over much of the role religions had played in aid and development.

In recent years, however, the religions have once again taken over much aid and development work. For over 50 years Christian aid agencies have worked on development issues, and over the last 25 years Islamic aid agencies have undertaken a similar role. This has gone even further in the last decade. Some countries, unable to cope with the burden, have handed back education and health to the religions. In Zambia, for example, 60 percent of all schools and over 40 percent of all health care is now run by the churches and mosques.

In addition to the Islamic non-governmental organizations covered here, Islamic states also provide funding, estimated as an impressive and often unacknowledged 30 percent of worldwide contributions to aid and development.

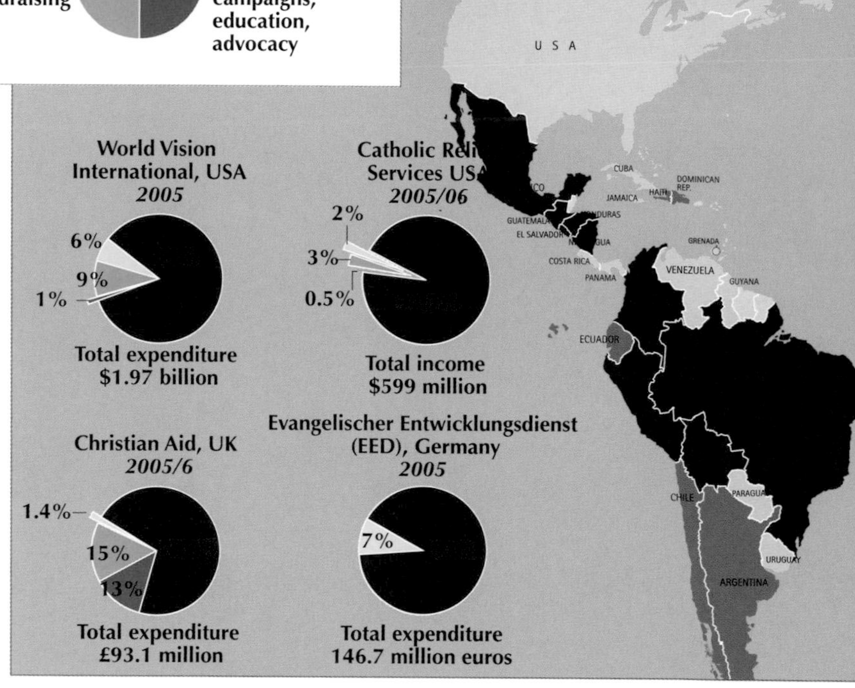

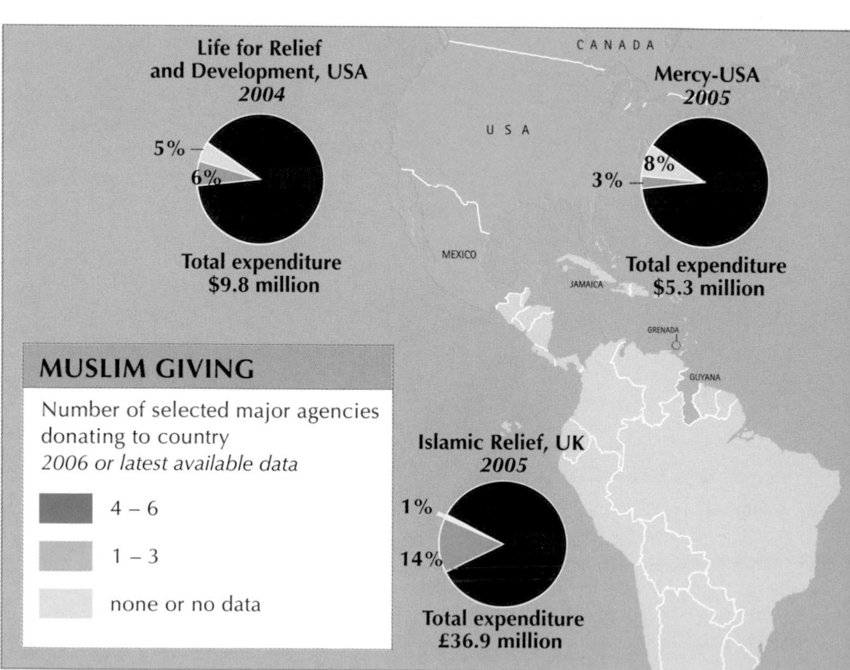

Aid and Development

Religious organizations make a substantial and growing contribution to aid and development work throughout the world.

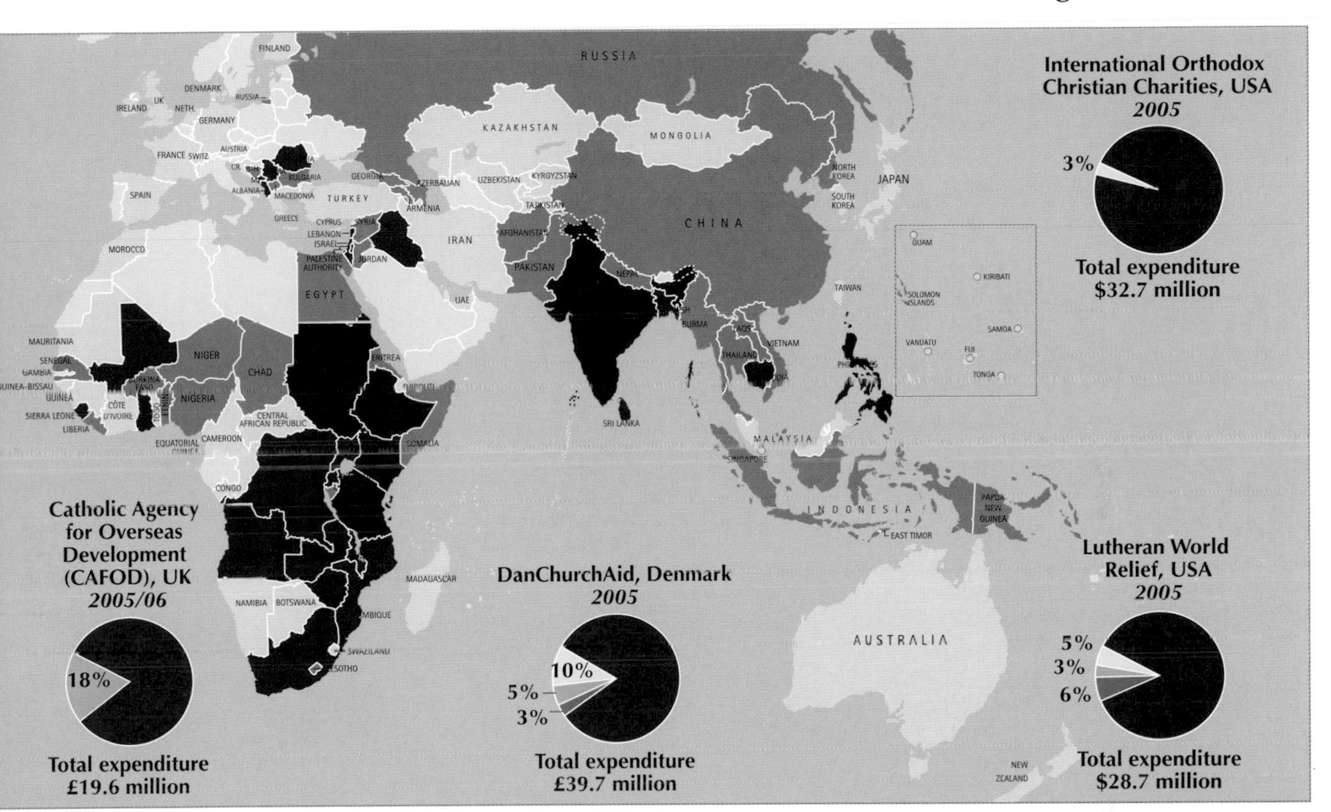

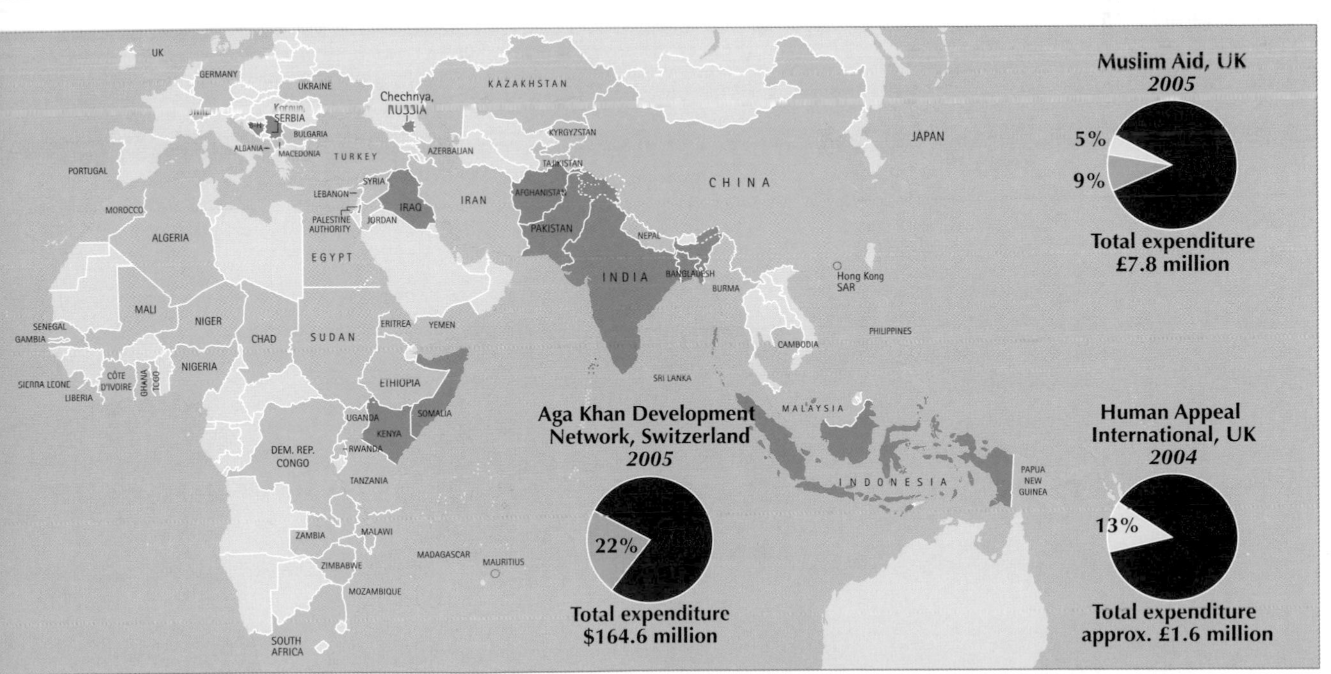

LEGAL SYSTEM

in Muslim majority states
2006

- Islamic law only, based on Shari'ah
- combination of Islamic and secular law
- Islamic law and secular law exist side by side
- secular law only
- non-Muslim states

Shari'ah, Islamic law, is founded upon the Qur'an, seen as the infallible word of God, and upon the Hadith, the sayings, actions and silent approval of the Prophet Muhammad. The formal term Shari'ah is Arabic for 'what is prescribed', and indicates correct ways of behaviour covering religious, political, social, domestic and private life. A further source of legal rulings is *ijma*, legal agreement, which refers to a method of framing laws based on consensus between jurists.

All Sunni Muslims belong to one of the four schools of classical law: Maliki, Shafi, Hanafi or Hanbali. Within Shi'a Islam, which does not follow any of these Sunni schools, the *ulama* or council of the 12 Shi'a elders, is the central political and legal institution. The judges appointed by the *ulam*a eventually became known as ayatollahs and formed their own courts.

The application of Shari'ah by Muslim majority countries varies according to the school of law and the extent to which the legal structure incorporates secular laws and customary codes. Senegal, for example, has a legal system based on secular French civil law, while Saudi Arabia enforces Shari'ah according to the strict Hanbali school of law. As a result of colonial influence, many countries have a hybrid system, combining secular Western legal interpretations and structures with elements of Shari'ah.

The *sadad* courts of Morocco deal with matters related to Muslim and Jewish personal law and religious jurisprudence. Appeals involving Shari'ah law can also be heard in Morocco's regional courts.

Shari'ah is the basis for legislation that is unique to Libyan socialism. People's Assemblies established by President Qadhafi deal with modern issues Shari'ah may not cover.

There are no Shari'ah courts in Tunisia, but the personal affairs of practising Muslims are governed by Shari'ah.

Shari'ah courts are integrated into the national court system in Egypt. The constitution recognizes the principles of Islamic jurisprudence as the primary source of legislation.

A framework of Shari'ah law is in force in northern Sudan but, under the peace agreement negotiated in 2005, legislation in southern Sudan is based on the traditional laws, religious beliefs, and customary practices of the people.

ALBANIA
TUNISIA
MOROCCO
ALGERIA
LIBYA
SENEGAL
GAMBIA
MALI
NIGER
CHAD
NIGERIA

Nigeria is officially a secular federation, although 12 northern states with Muslim majorities have now adopted parts of the Shari'ah code.

Mali's civil law is based on the French legal code, and family law is based on Shari'ah and customary law.

Islamic Law

Shari'ah sets out for Muslims the basic legal and ethical codes upon which all relationships and institutions should be founded and by which they should be guided.

Although Turkey is more than 97% Muslim, its legal system draws on Swiss, Italian, French and German civil codes that were introduced in 1926 following the establishment of a Turkish national state in 1923.

Syria has separate legal systems for civil, criminal and personal status matters. There are also separate courts for Muslims, Druze, Christians and Jews in relation to family law.

In Kuwait, in the absence of Shari'ah provision for legal cases, judgements are made according to customary law.

Afghanistan is developing a legal framework based on Shari'ah, customary law and Islam-guided state legislation.

Pakistan combines Shari'ah and English common law, but the constitution states that all laws should be in keeping with the Qur'an and the Sunna.

Iran combines Shari'ah, constitutional law and customary law, but all laws must be based on Islamic criteria. The power of the *ulama* is based on the mullahs, priest teachers who transmit decisions and instructions throughout the community.

Although secular laws exist in UAE, Qatar, Bahrain and Yemen, their constitutions declare Shari'ah as the principal source of legislation.

Indonesia's legal system is based on Shari'ah, the Dutch civil code and customary law. In 2003, Aceh became the only province in Indonesia allowed to establish full Shari'ah courts.

The Sultan of Oman is the final authority on the Shari'ah Ibadite legal code and state legislation. The Ibadiyyah tradition originated before the split between the Sunni and Shi'a.

The Shari'ah Hanbali legal code is strictly enforced in Saudi Arabia for family and criminal law. Commercial law is governed by royal decrees and codes.

The Malaysian legal system combines Shari'ah and English common law. Family law relating to non-Muslims comes under federal jurisdiction.

The civil code in Iraq recognizes Shari'ah as a formal source of law in matters of personal status.

HAMEDIAN GALLERY
SPECIALIZED IN OLD RUSSIAN ICONS
OLD ORIENTAL CARPETS, OLD JEWELLERY
& ANTIQUES
ויאה דולורוזה
طريق الآلام
VIA D LOROSA

Part Four CONFLICTS & TENSIONS

One of the most common charges against religions is that they have been the source of more violence than peace, and that the world would be a better place without them and their rivalries.

There is some truth in this. Religious divisions – faultlines – run across continents and through time and still affect politics, economics and communities to this day. Historical arguments, wars, struggles and internal disputes have created the contemporary maps of the world in ways that few appreciate. For example, the 'Christian' EU arises from the experiences of invasion by Islam from the 14th to 17th centuries, and the occupation of part of Eastern Europe by Islam right up to the 20th century. The violent conflicts in Iraq have their roots in the split between Sunni and Shi'a Islam in the 7th century, while struggles in Sudan, Ethiopia and Nigeria can be traced back in some areas to the 10th century.

However, the violence has not only been from the religious side. In the last 100 years, the major religions have been more heavily persecuted than at any other time in history. And most of this has not been religion persecuting religion. It has been ideologies persecuting religion. This ranges from the Mexican socialist revolution in 1924 attacking the power, land holdings and ultimately the clergy and buildings of the Catholic Church, through the attacks on all faiths in the Soviet Union, the Holocaust of the Jews by the Nazis, the massive onslaught of the Chinese Cultural Revolution against all faiths in the 1960s and the assaults on the Baha'is in Iran from the 1970s onwards.

Sadly, the faultlines persist: as the religions emerge from persecution, some are starting again with their own persecutions. However, for many religious groups, time and the experiences of the past 100 years, plus the impact of ecumenical and interfaith movements, have begun to change them, and old divisions and enmities are fading.

An Arab passing Israeli soldiers on the Via Dolorosa, Jerusalem. The Via Dolorosa is the path that Christ took on his way to the Crucifixion

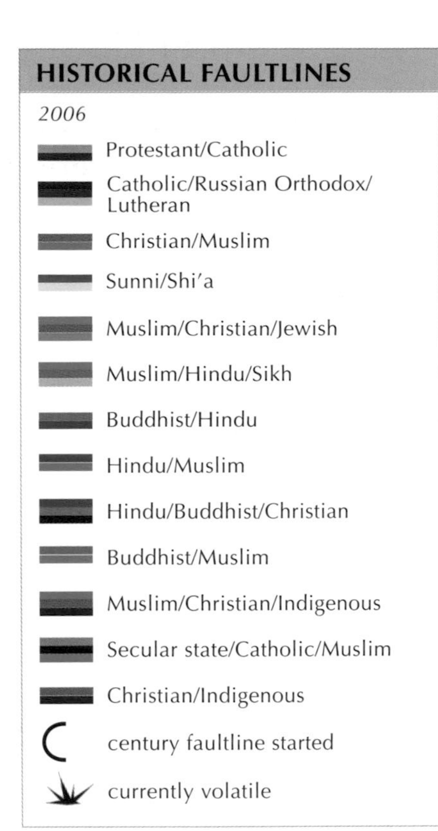

Historic faultlines, many over a hundred years old, created by differences in religious belief and practice, lie at the root of many of today's areas of tension and struggle. In Iraq, the split between Sunni and Shi'a Muslims, dating from the late 7th century, fuels the civil strife which has so affected the country since the fall of Saddam Hussein. The faultline between Islam and Christianity in Eastern Europe and the Caucasus is illustrated by the controversial application of Turkey to join the European Union. The split between Catholic, Russian Orthodox and Lutheran Churches still has repercussions in Europe and Russia.

Some faultlines have become regions of cultural difference rather than sources of major tension, such as the Swahili coastline of East Africa. Other ancient faultlines, such as those between Indonesian Christians, Hindus and Muslims, have been re-awoken where only a few years ago communities lived side by side.

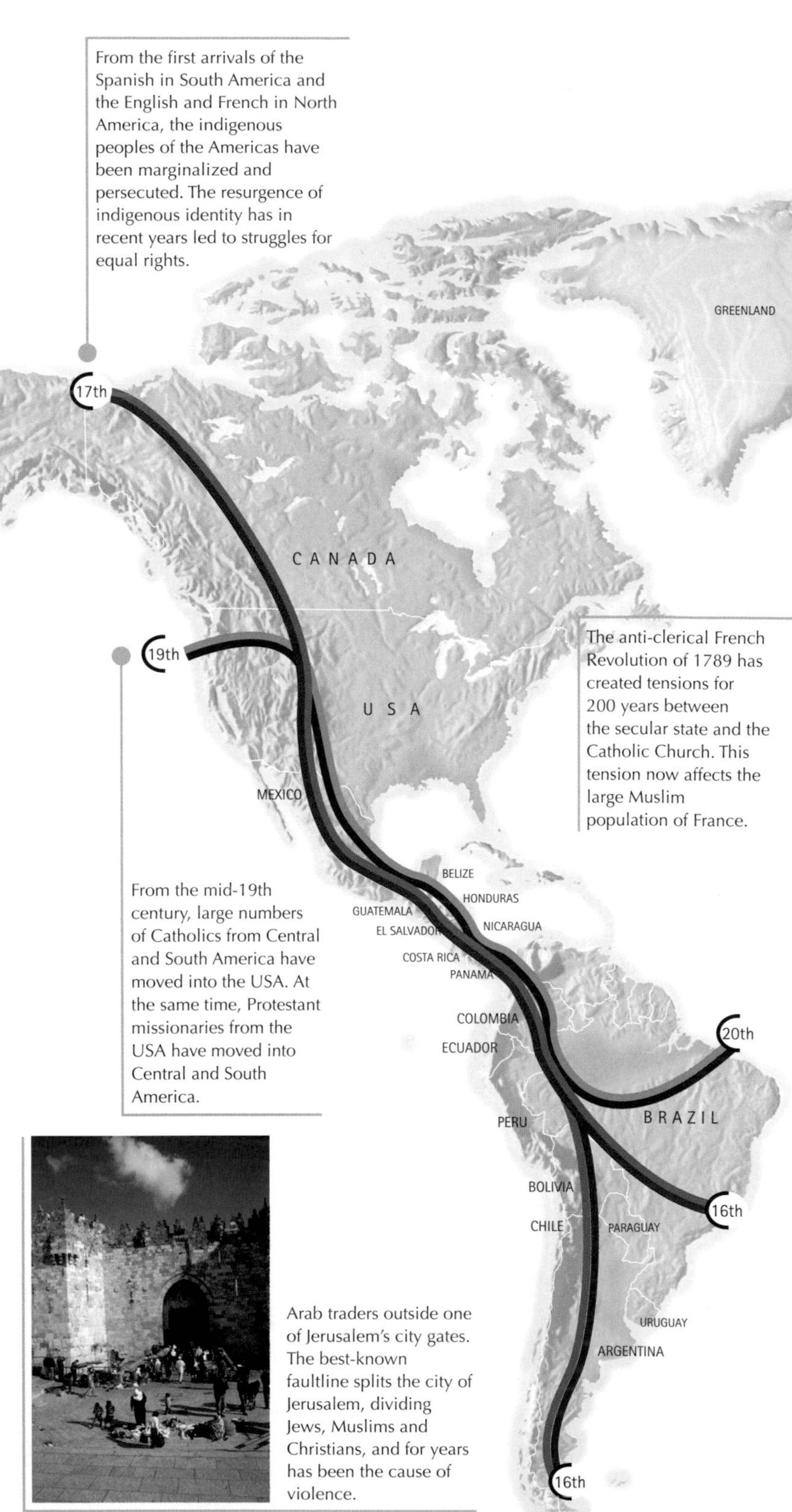

Arab traders outside one of Jerusalem's city gates. The best-known faultline splits the city of Jerusalem, dividing Jews, Muslims and Christians, and for years has been the cause of violence.

Faultlines

Historical feuds and struggles between religions still shape many contemporary events around the world.

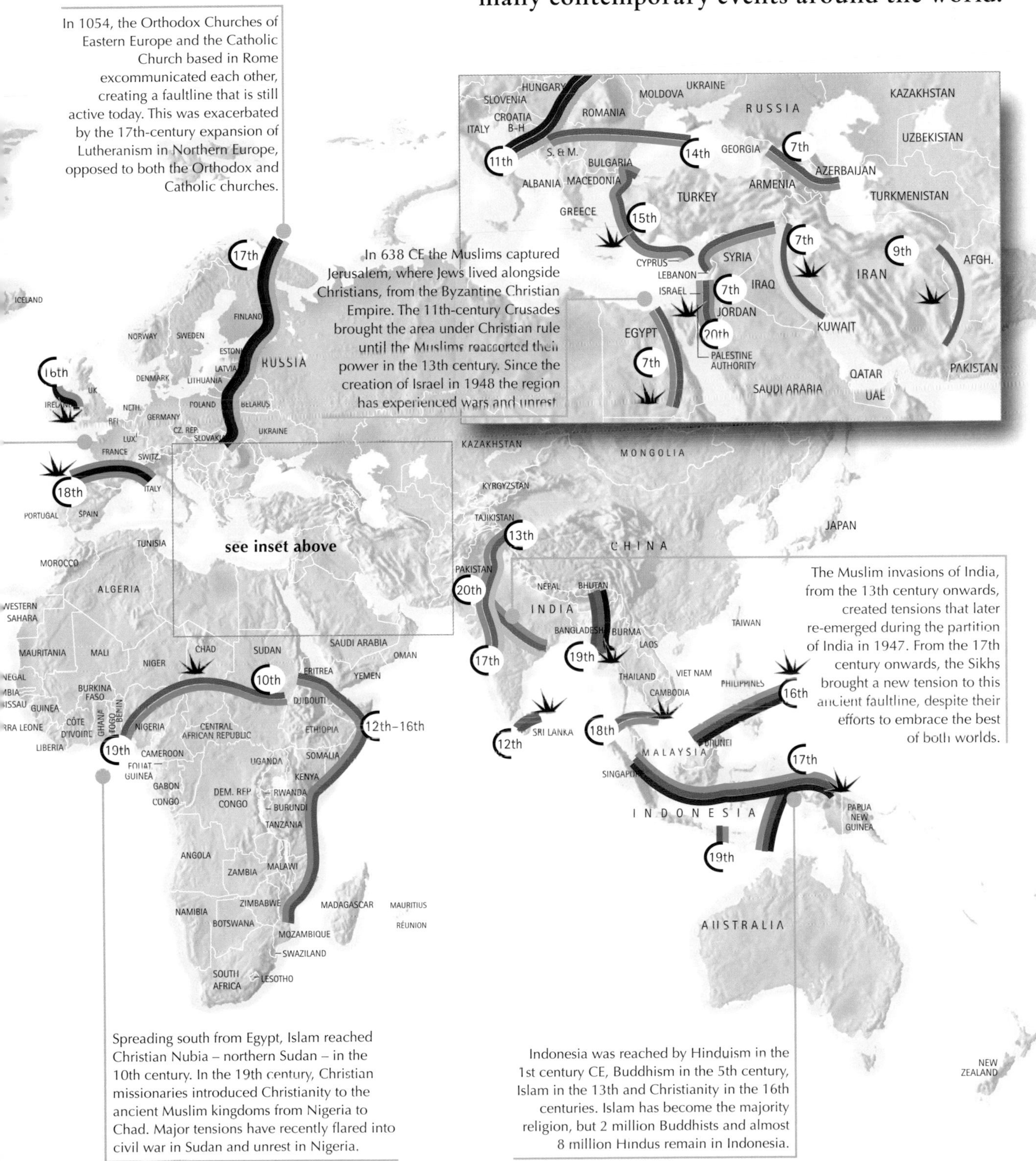

In 1054, the Orthodox Churches of Eastern Europe and the Catholic Church based in Rome excommunicated each other, creating a faultline that is still active today. This was exacerbated by the 17th-century expansion of Lutheranism in Northern Europe, opposed to both the Orthodox and Catholic churches.

In 638 CE the Muslims captured Jerusalem, where Jews lived alongside Christians, from the Byzantine Christian Empire. The 11th-century Crusades brought the area under Christian rule until the Muslims reasserted their power in the 13th century. Since the creation of Israel in 1948 the region has experienced wars and unrest.

The Muslim invasions of India, from the 13th century onwards, created tensions that later re-emerged during the partition of India in 1947. From the 17th century onwards, the Sikhs brought a new tension to this ancient faultline, despite their efforts to embrace the best of both worlds.

Spreading south from Egypt, Islam reached Christian Nubia – northern Sudan – in the 10th century. In the 19th century, Christian missionaries introduced Christianity to the ancient Muslim kingdoms from Nigeria to Chad. Major tensions have recently flared into civil war in Sudan and unrest in Nigeria.

Indonesia was reached by Hinduism in the 1st century CE, Buddhism in the 5th century, Islam in the 13th and Christianity in the 16th centuries. Islam has become the majority religion, but 2 million Buddhists and almost 8 million Hindus remain in Indonesia.

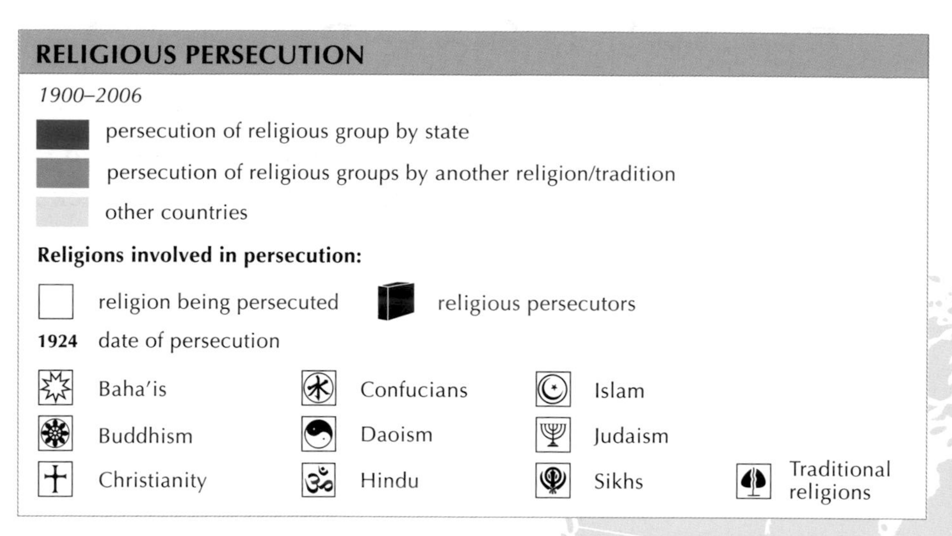

Religion is often criticized on the grounds that most wars arise from religious tensions and disagreements. While this might have been true in centuries past – although highly disputable – it has certainly not been the case in the last 100 years. Instead, religion has been violently persecuted by secular ideologies. Communism, fascism, socialism and nationalism have all seen religion as the major threat to their vision of creating new societies, because in many cases religion was a major prop of the state that the revolutionaries sought to overthrow. As a consequence, an unprecedented assault on religious buildings, personnel and ordinary believers has taken place.

However, in recent years, with the collapse of these ideologies and the recovery of many religions in different parts of the world, there has been a marked rise in religious-based attacks, violence and warfare.

Marxism stated that with the coming of socialism and communism, 'religion would whither and die'. In fact, the reverse has proved true. It is the ideologies that have withered and died, albeit at the cost of tens of millions of lives. Religion has not only survived but has often been the inspiration for the resistance movements that helped to overthrow the ideologies. From the Catholic Church in Poland, through the Lutherans of East Germany, to the Buddhists of Cambodia and the Muslims of Central Asia, religion has outlived the power of coercive regimes. While no religion has quite the role it had before such massive social changes and persecutions, nevertheless religion has returned to centre stage in many countries to play its part once again in building up and sustaining nations, peoples and cultures.

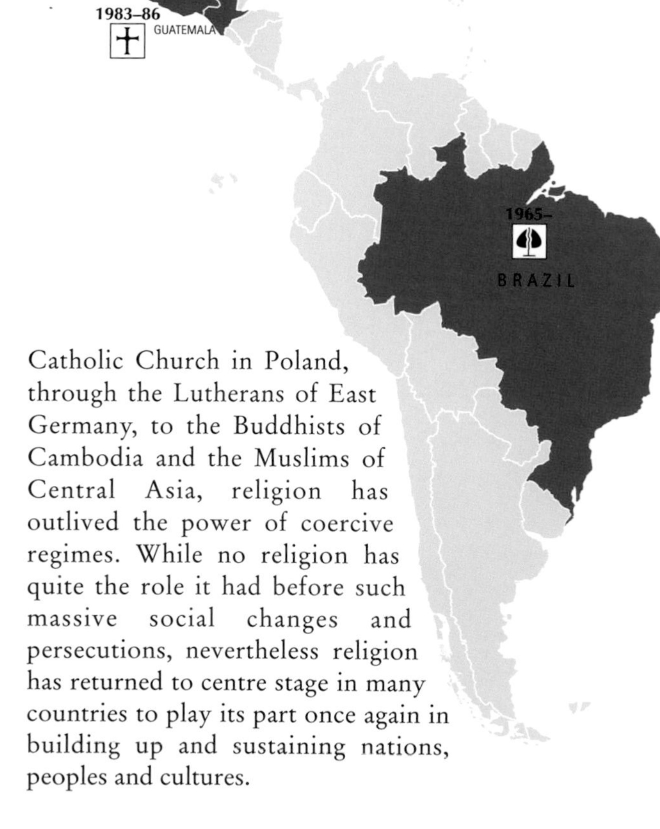

Emerging from Persecution

Despite a century of more intense persecution than ever before, the major religions have survived and are growing again, where once they were attacked.

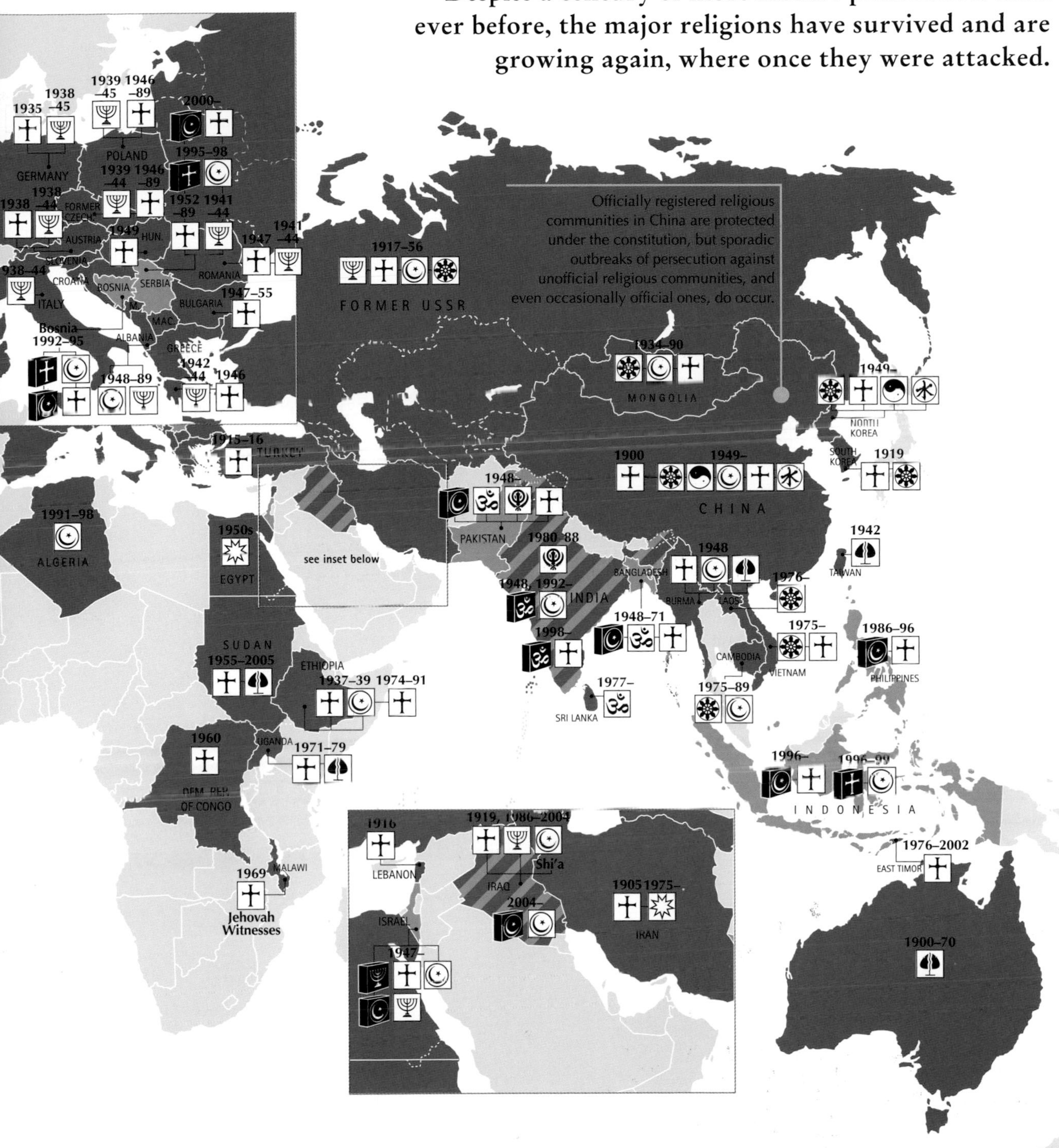

WHEREFORE TAKE
UNTO YOU THE
WHOLE ARMOUR
GOD, AND YOU
MAY BE ABLE TO
WITHSTAND IN THE
EVIL DAY, AND
HAVING DONE ALL
TO STAND.
EPHESIANS 6:13

Part Five CONTEMPORARY CHALLENGES

The major religions survive by appearing to be timeless, while actually adapting. In the last 100 years, they have had to respond to profound changes in social behaviour.

One of the biggest challenges has been the demand that women be treated as equal with men. The religions have generally been less responsive to this than other institutions, with Islam and Catholicism refusing to give ground on issues concerning authority and clerical power. Indeed, most religions are bastions of patriarchal structures and are proving resistant to the rise of women's rights, although changes have begun, not least in Judaism and the Anglican Church.

An area where the religions have embraced new ideas is that of ethical investment. Beginning with Islam, which built upon its ban on usury, all major religions are now considering moving their vast reserves into ethical investments. The centrally held funds of the key religions are estimated to total $7 trillion, which the religions now see as offering the potential for social good.

Religions have also become active is the environment. Drawn in by groups such as the World Wide Fund for Nature (WWF), and its sister group the Alliance of Religions for Conservation (ARC), all major religions now have extensive environment programmes, ranging from eco-audits of their buildings, through ecological management of their forests and land holdings, to education and advocacy work.

The threat of HIV/AIDS has cast a shadow on the religions. Initially, many saw it as divine punishment for the sexual activities of homosexuals. However, a sense of realism, helped by the campaigns of groups such as the WHO and UNICEF to win the religions over to the side of prevention and care, has led to a massive shift in religious opinion. Now, most religions are actively involved in such work at all levels.

One of the encouraging developments in the last 20 years or so has been the growth of an active and socially engaged interfaith movement. Increasingly, religions are coming together to tackle social and religious issues of contention. This offers some hope for the future, which, although it threatens a continued rise in religious tension, is also likely to be increasingly pluralistic.

Christians gathered before the US Capitol in Washington DC on a March for Jesus rally

WORKING TOGETHER

2006

Churches involved in HIV/AIDS work

Interfaith activities

social action and interfaith dialogue

interfaith dialogue only

Since the first gathering of key Protestant Christian traditions in 1910, the ecumenical movement has grown to embrace both the Catholic and Orthodox Churches. Some denominations have reunited after centuries apart and a long history of mutual antagonism, although recently the impetus has slowed down and even stalled over issues such as women priests and gay rights. In many countries, the Churches work side by side to tackle social issues, including HIV/AIDS prevention and treatment, for which they contribute around 40 percent of all palliative care.

The roots of interfaith dialogue lie in the efforts of Hindu leaders to reach out to the world's religions from the late 19th century onwards. Since the 1950s, almost all religious leaders and religious communities have engaged in such dialogue, even if much of it is at a formal and often superficial level.

A century of massive persecution by non-religious forces has engendered a shared concern between religious communities that they will be prevented from following their chosen religion. They have also been drawn together by their response to crises such as inter-communal violence, natural disasters and HIV/AIDS. This has led to religious leaders being far less likely to be antagonists than to be working collaboratively – or at least side by side – to alleviate suffering.

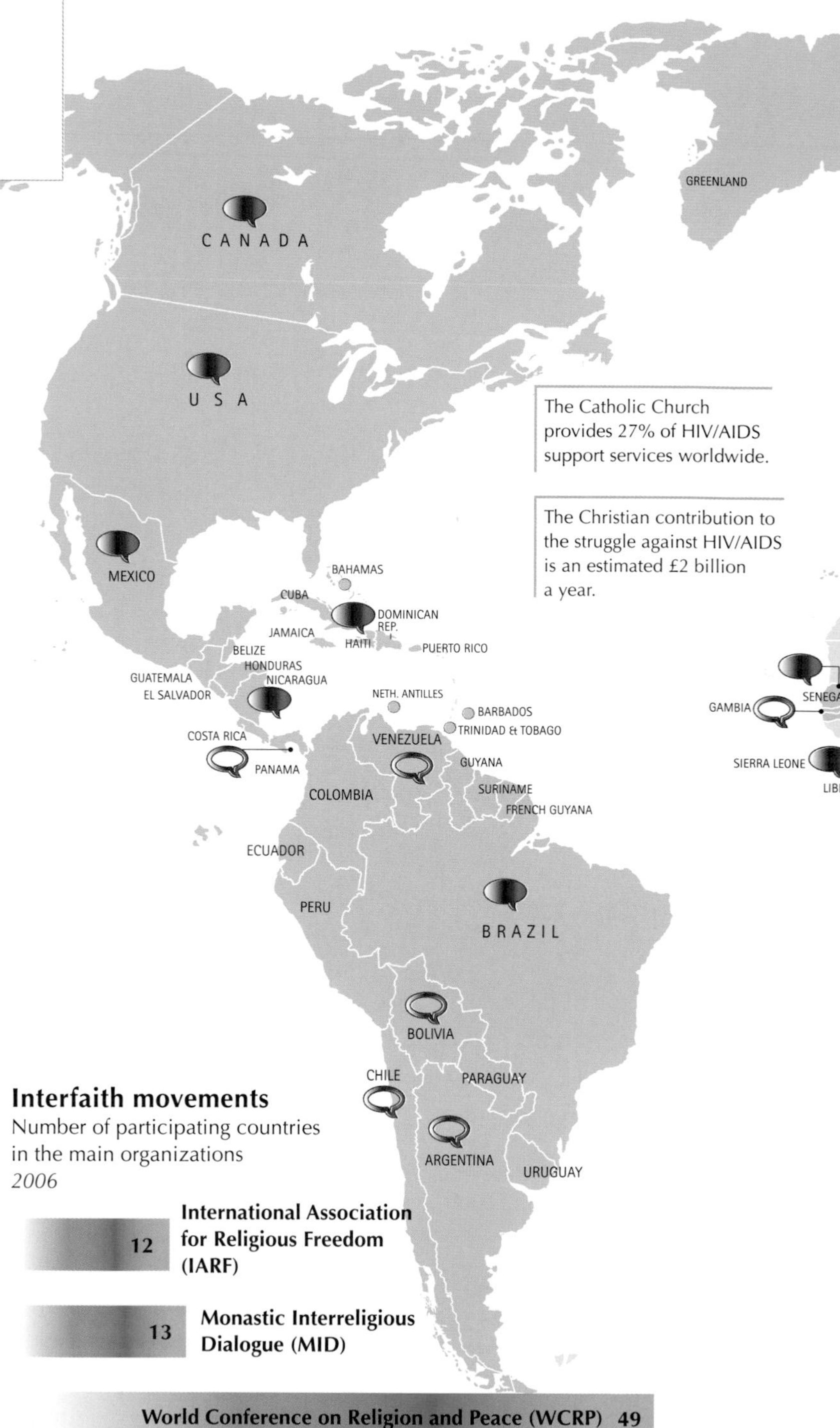

Shared World

Ecumenical Christian and interfaith movements are growing stronger. They are increasingly addressing social, health, environmental and human rights issues.

In 2004, 97% of churches in Kenya, Malawi, Mozambique, Namibia, Swaziland and Uganda were caring for HIV/AIDS orphans.

50% of all medical facilities in Zambia are run by the Churches.

The Council for a Parliament of the World's Religions, the world's first major interfaith event, took place in 1893 in Chicago, USA. It was revived for its centenary in 1993, when around 8,000 delegates of all faiths gathered in Chicago. A similar number, from over 70 countries, have met at subsequent events in South Africa and Barcelona to deepen inter-faith understanding and explore spiritual responses to critical issues such as HIV/AIDS, international debt, access to clean water, environmental degradation and conflict resolution.

STATUS OF WOMEN

Within organization of majority religion/tradition *2006*

- equal at all levels
- equal at some levels
- subordinate

Direction of trend:

- moving towards equality
- increasingly restricted

The change in the social status of women in many cultures over the last 50 years has confronted all the major religions with profound challenges. Within Christianity it has led the Lutherans and Anglicans, and within Judaism the Reform and Conservative branches, to include women within the ranks of the clergy. It has also provoked strong statements against an equal role for women from the Catholic Church and Orthodox Judaism. Neither is there acceptance of the idea of women achieving equal status within the Islamic religious hierarchy, although some traditions allow women to lead prayers for women-only congregations, and there have been instances of women leading mixed congregations.

Women's role in Buddhist ceremony has been under debate, often led by Western or Westernized Buddhists, who have called for equal rights for women to be monastic in countries such as Thailand, where the tradition has been for male monastics.

Sikhism is built upon teaching that men and women are equal – although this is not always borne out at the priestly level. Hinduism and Jainism, however, have continued to exclude women from most positions of authority, except within monastic orders for the Jains. In Hinduism, some major female swamis and gurus are accepted on the strength of their charismatic personalities.

Deep-seated social attitudes towards women often have more influence than theology on religious attitudes towards women. In the Anglican Communion, where women are officially counted as equal, the 'stained-glass ceiling' effectively means they are denied full equality and are often the target of prejudice and subtle discrimination.

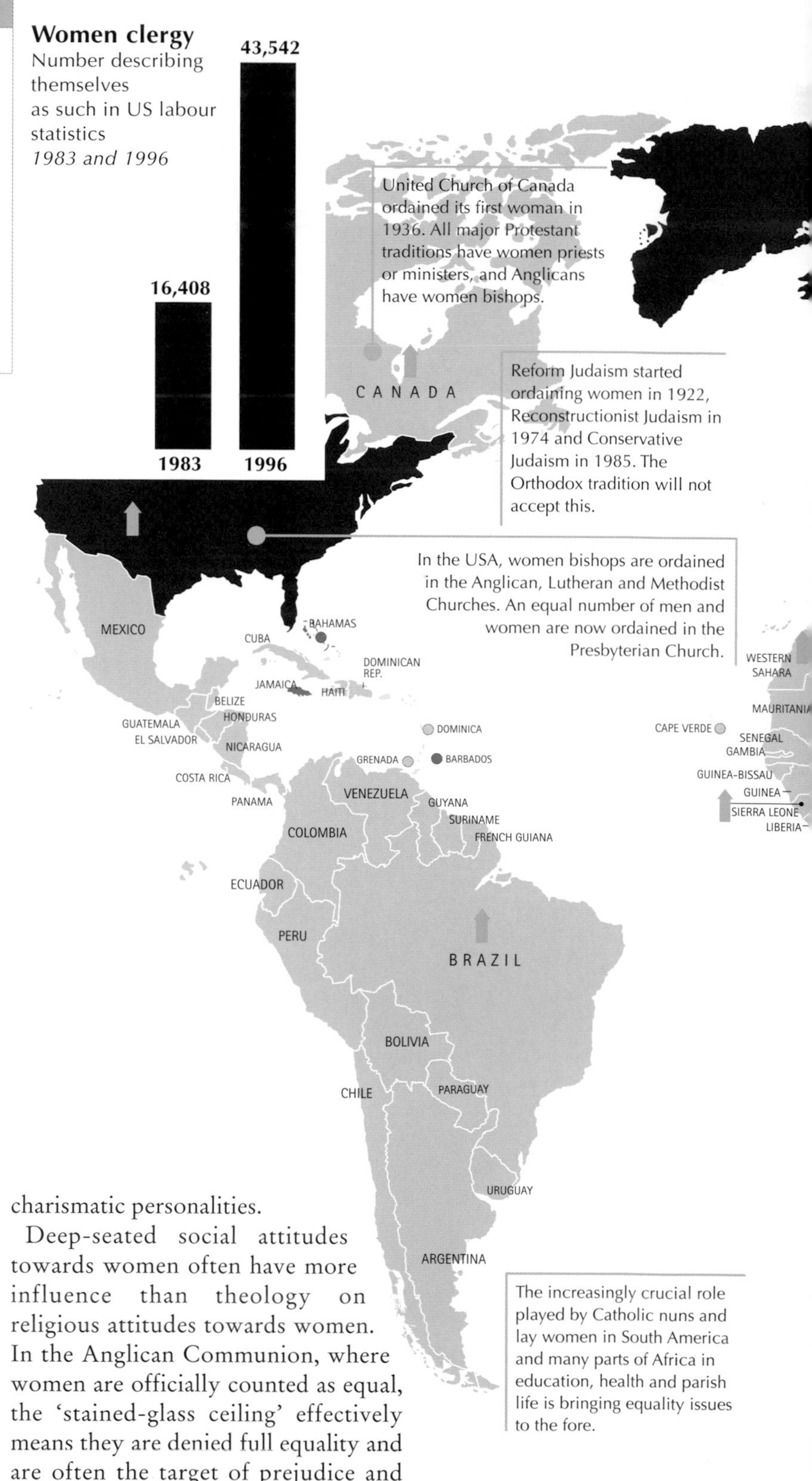

Equal Rites

Within some religious organizations women have equal status with men when it comes to administering religious rites. Other religions are firmly opposed to women fulfilling this role.

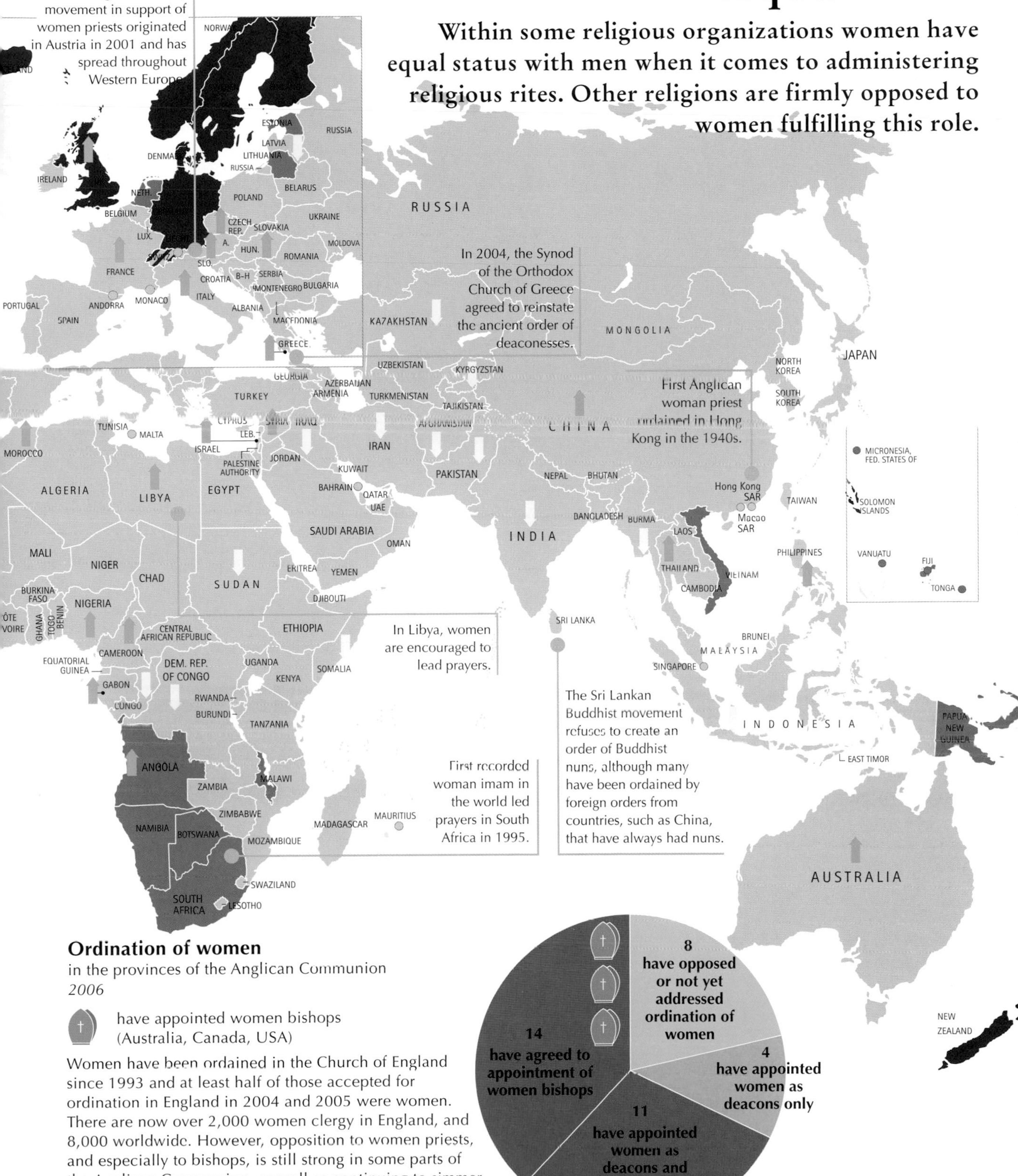

Ordination of women

in the provinces of the Anglican Communion

2006

have appointed women bishops (Australia, Canada, USA)

Women have been ordained in the Church of England since 1993 and at least half of those accepted for ordination in England in 2004 and 2005 were women. There are now over 2,000 women clergy in England, and 8,000 worldwide. However, opposition to women priests, and especially to bishops, is still strong in some parts of the Anglican Communion, as well as continuing to simmer within those Churches that have ordained women.

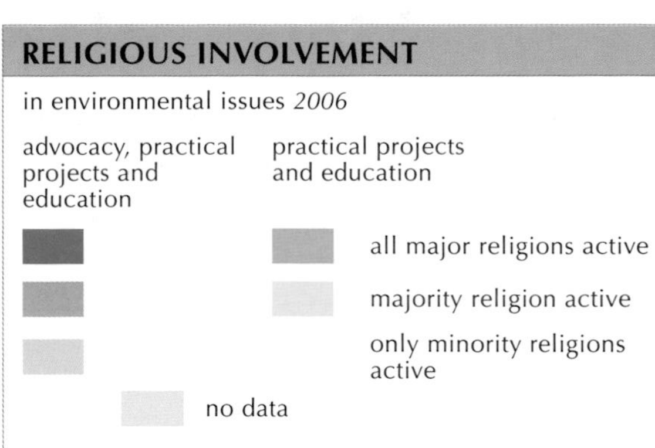

The major religions of the world own about 7 percent of the habitable surface of the world and are among the largest and most powerful investors. In recent years they have begun to explore ways in which their land – forests, farms and urban sites – can be managed in an environmentally friendly way. They are also seeking to move their investments into funds that use the environment sustainably. Across the world, religions are now among the most active agencies in environmental management.

The religions have also explored their own teachings and traditional practices and discovered that these often contain insights into relating to, and using, nature. For example, Islam has legal codes for environmental protection, dating from the 9th century CE, while the traditional way Daoists in China have managed their sacred mountains has preserved these environmentally important areas.

The religions also have a major role as advocates and as voices of authority within many cultures. In the USA, all major religions have combined to work on changing the government's approach to climate change issues, while in many parts of Africa the Churches' position of trust enables them to persuade local people to adopt more eco-friendly lifestyles.

Increasingly, new partnerships between the religions and environmental and development organizations are becoming the norm in many areas of the world, marking a major shift in both religious and secular attitudes towards each other, driven by the realities of the environmental problems that we now face.

CANADA

USA

MEXICO

CUBA

JAMAICA

HAITI

DOMINICAN REP.

BELIZE

HONDURAS

GUATEMALA

EL SALVADOR

NICARAGUA

COSTA RICA

PANAMA

VENEZUELA

TRINIDAD & TOBAGO

GUYANA

SURINAME

FRENCH GUIANA

COLOMBIA

ECUADOR

PERU

BRAZIL

BOLIVIA

CHILE

PARAGUAY

URUGUAY

ARGENTINA

GREENLAND

SIERRA LEONE

Catholic bishops in Canada and the USA have issued Pastoral Letters to safeguard the Colombia River, and involved all major stakeholders in a five-year investigation of its condition.

A project in Mexico is helping conserve plants such as bromeliads, orchids and cycads, harvested for use in traditional celebrations and religious practices. It is using the knowledge of indigenous people to develop strategies for managing the resources of the rainforest in a sustainable way.

Many Christian groups throughout Brazil are actively involved in urban and rural environmental work. Amongst them, the Benedictine community run forest and water restoration projects and the organic cultivation of medicinal, fruit and vegetable crops.

Environmental Protection

Religious organizations are the largest social network in the world. In recent years they have turned their attention to caring for the environment.

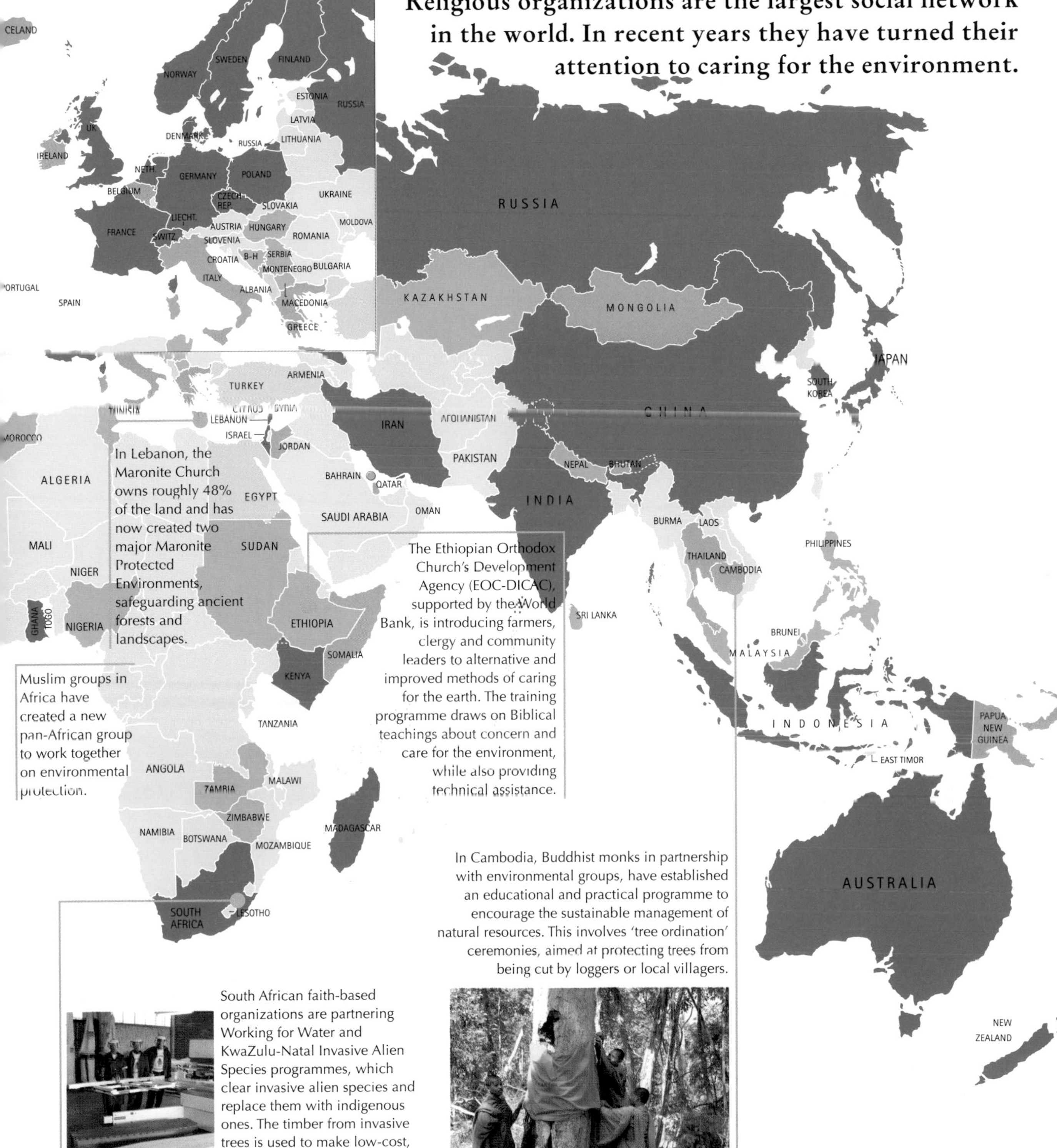

In Lebanon, the Maronite Church owns roughly 48% of the land and has now created two major Maronite Protected Environments, safeguarding ancient forests and landscapes.

The Ethiopian Orthodox Church's Development Agency (EOC-DICAC), supported by the World Bank, is introducing farmers, clergy and community leaders to alternative and improved methods of caring for the earth. The training programme draws on Biblical teachings about concern and care for the environment, while also providing technical assistance.

Muslim groups in Africa have created a new pan-African group to work together on environmental protection.

In Cambodia, Buddhist monks in partnership with environmental groups, have established an educational and practical programme to encourage the sustainable management of natural resources. This involves 'tree ordination' ceremonies, aimed at protecting trees from being cut by loggers or local villagers.

South African faith-based organizations are partnering Working for Water and KwaZulu-Natal Invasive Alien Species programmes, which clear invasive alien species and replace them with indigenous ones. The timber from invasive trees is used to make low-cost, high-quality Eco-coffins, in which the poor may bury their dead with dignity.

ETHICAL BANKING WORLDWIDE

Involvement of religions in ethical investment
2006

Muslim banks operating according to Islamic principles

majority religion supports ethical banking through advocacy, campaigning or investments

other countries

international centre of religiously orientated ethical investment

The rise of the religious ethical investment movement is of considerable economic importance: the combined investments of the major religions are estimated at $7 trillion.

The ethical investment movement began in the 1960s during the struggle for civil rights in the USA. Churches opposed to the apartheid system in South Africa withdrew certain investments. In 1974 the first purely Islamic bank opened, offering an alternative to conventional banking that was based on the Qur'anic ban on the taking and charging of interest.

By 2000, the emphasis within the Churches was to combine disinvestment with active investment in socially responsible companies, which produced a good return, supported employment rights and were environmentally sustainable. This was known collectively as the 'triple bottom line'.

In 2005, 3iG, the International Interfaith Investment Group, was founded, bringing together finance managers from many different religions and linking Buddhists, Christians, Daoists, Hindus, Jews, Muslims and Sikhs. Meanwhile, Islamic banking has become so significant that all the major secular banks offer Islamic banking facilities.

Ethical Investment

Worldwide, the major religions are moving their financial assets into socially responsible investments – ranging from Islamic banking to ethical investment portfolios.

Daoism, a major element in Chinese traditional religion, bans usury and supports ethical investment.

Worldwide, over 300 Islamic financial institutes have assets estimated at $250 billion. These are expected to increase by an annual 15%.

In 2001, religious communities and their members held 38% of the global share of the international investment portfolio.

Muslim traders in a street market in Pakistan. In Islam, each financial or commercial transaction is formally blessed and sealed by the word 'Bismallah', which invokes God's witness of the integrity of the deal.

THE DIVERSITY OF RELIGION

Predicted changes to major religions and religious belief to 2025
2006

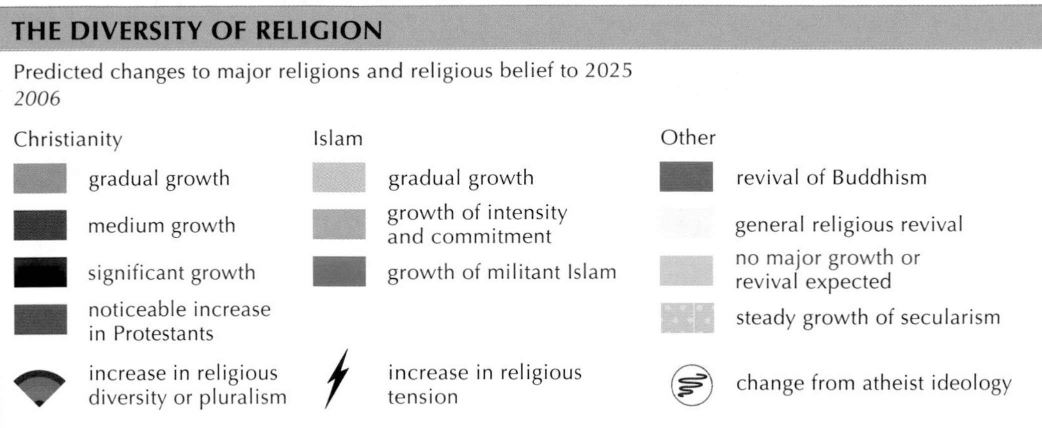

The continuing revival of religion is one of the clearest indicators for the future. In the first edition of this atlas, we represented religion emerging from the newly liberated communist countries. We also forecast the rise of militant Islam, which has indeed become a force to be reckoned with. It has in turn sparked reaction from other religions as well as military responses.

We see the continued growth of pluralism worldwide – especially in those areas where communism has either withered away or is fading – such as Mongolia and China. The rise of diverse and new religious communities in the vacuum left behind is a marked feature of such societies.

The experience of migrant workers from Eastern Europe, working in major, multi-faith cities such as Paris and London, will affect their home countries in two ways: firstly, by introducing religious diversity to countries where the traditional religions have long held sway, and secondly by secularizing many of the young. However, the rise of secularism will continue to be a European exception to the picture elsewhere in the world.

The future will also continue to be affected by history. The Faultlines map (pp64–65) should perhaps be read in conjunction with this one, to indicate potential areas of conflict and stress.

The Future

Religion will respond in a diversity of ways to both the opportunities and challenges raised by an increasingly free market world.

Part Six HEARTLANDS

In the beginning, over 4,000 years ago, religions emerged locally and stayed local. As far as we can see, no Neolithic religious figure from the long barrows and ancestor religion of Britain went on a missionary journey to Egypt to try and convert the Pharaohs, and no priest of Isis went from Egypt to Babylon to try and convert the Babylonian priests of the ziggurats. This tradition of local, indigenous religions is the one that is gradually dying out or being absorbed by the missionary religions, which began with Buddhism and developed through Judaism, Christianity and Islam. They are now being replicated in the moving out from their original heartlands of major religions such as Hinduism, Jainism, Daoism and the Baha'i faith.

It is only when a religion becomes global, or at the least intercontinental, that its heartlands become truly significant. The urge to go on pilgrimage, to see and venerate the holy places associated with the founder or the most momentous events in the faith's history, comes when those places are no longer just around the corner.

The major religions are creations of, and spring out of, the Old World – the world of antiquity from Europe, through the Middle East to Asia. This is where the great religions have their heartlands – the places that provide the setting for the stories and personalities that have shaped the major religious forces of the world.

One way in which heartlands and sacred places are being appreciated is for the astonishing role they play in helping to preserve and conserve some of the most beautiful and wild places on the planet. In a joint study, WWF and ARC charted the extent of many of the world's most important national parks, world heritage sites and important conservation areas. They have survived this long precisely because they are places held sacred for centuries, or indeed millennia. Their holiness has led people to treat with respect the areas around sacred mountains, valleys, rivers and even cities. Logging, hunting or development have often been more effectively banned and blocked because of the sacred nature of the place, rather than because of official government edicts or laws.

In addition to the physical heartlands of the religions, there are the spiritual heartlands – the teachings, scriptures, practices and core beliefs that open windows onto the souls of the major faiths. In the end it is these that bestow upon the material world its significance upon the material world.

The Hanging Temple, northern Heng Shan Daoist Mountain, China

FOUNDATIONS AND DIVISIONS

Origins of the major religions
founders/leaders, dates and places

BCE Before Common Era (BC)
CE Common Era (AD)
not given after 1000 CE

- Baha'i
- Buddhism
- Christianity
- Confucianism
- Daoism
- Islam
- Jainism
- Judaism
- Sikhism

Division of the major religions
founders/leaders, dates and places

BCE Before Common Era (BC)
CE Common Era (AD)
not given after 1000 CE

- Buddhism
- Christianity
- Daoism
-  Hinduism
-  Islam
-  Judaism

The power of holy places, associated with the birth of a great religious figure, or a major centre of religious power, can be seen in the vast numbers that go on pilgrimage to such sites. Rome, Jerusalem, Makkah and Medina are examples of the incredible draw such places have.

In some religions, the founders are the main attraction. For example, in Confucianism, the birthplace of Kong Fu Zi has drawn pilgrims for over 2,000 years. In other religions, the place of spiritual and temporal authority – such as Rome within the Catholic Church or Kufa in Shi'a Islam – exercises a greater attraction. For yet other traditions, their place of origin – Bristol for the Methodists, Geneva for the Calvinists – is of less importance, but does provide a sense of history.

In a few cases, entire cities have arisen from the origins of a religion, as with Salt Lake City and the Mormons. Jerusalem has become a holy city for Jews, Christians and Muslims alike.

The 'Old World' origin of the major world religions is clearly indicated by the distribution of such places of origin. The 'New World' of the Americas, Oceania, Australasia, and almost all of Africa has no significant centres of origin, though all these areas are centres for the thriving development of new religious traditions and for the resurgence of traditional religions, often in connection with one or other of the major religions.

Joseph Smith/Brigham Young
The Church of Jesus Christ of Latter-day Saints (Mormonism)
1847
Salt Lake City

U S A

Origins

Places associated with the origins of a religion, or where major new expressions came to life, are often centres of pilgrimage and veneration.

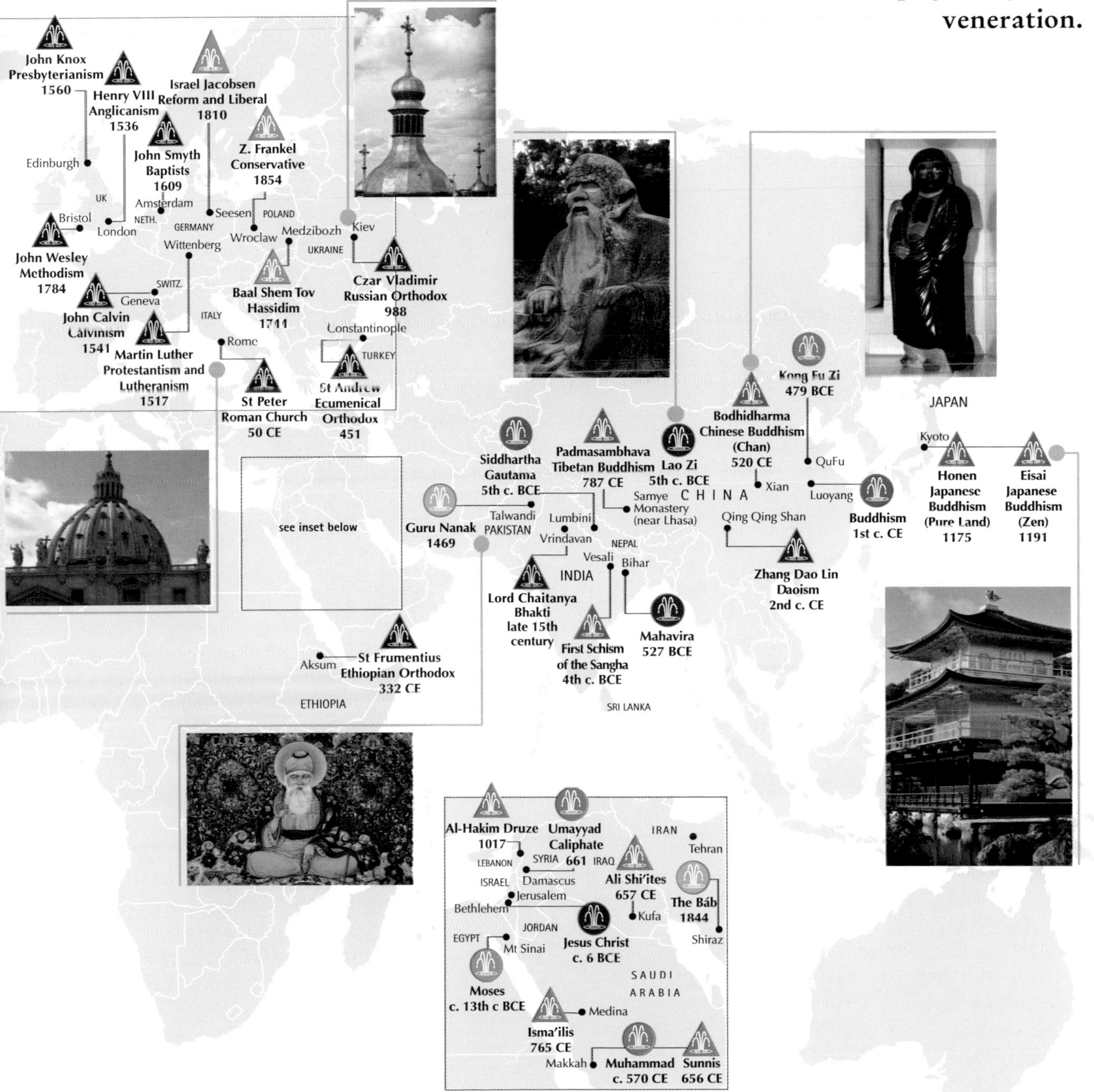

HOLY PLACES

Protected natural sites that are centres of devotion for major world religions as reported by WWF and ARC *2006*

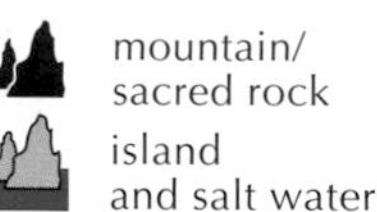

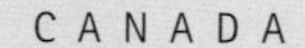

Kootenay National Park

Wupatki ruins, Arizona

Throughout the world, hills and valleys, rivers, springs and mountains have been venerated as sacred sites. From Hopi Indian ruins in Arizona, through the Lebanese Cedars of God, to holy mountains in China, the world has a sacred topography.

Today, many of these ancient sacred sites are important not just for their spiritual associations but because they are the centre of, or the reason for, the survival of wilderness areas. Sacred sites and landscapes have been protected from hunting, logging, farming or urban encroachment because they are holy places.

Recognition of this link has led international bodies concerned with wildernesses, national parks and conservation to work with the major religions and traditional local religions to map the relationship between the sacred and the natural. The partnerships being created build upon the extraordinary quality of sacred sites that are also places of significant bio-diversity and natural beauty.

Holy Natural

For many religions, nature offers a place of encounter with the Divine, helping to conserve some of the most beautiful places in the world.

Laponian area
Traditional

Pyätunturi Nat. Park
Traditional

SWEDEN

FINLAND

Northern Karelia
Christianity

ESTONIA

Lake Puhajärv
Buddhism, Christianity

RUSSIA

Yukanskiy Kanthy
Christianity

Khovsgol Lake
Buddhism

Bogd Khan Mt
Buddhism

Vangiin Tsagaan Uul
Buddhism

MONGOLIA

Ierelj Nat. Park
Buddhism

Mount Tai Shan, China

Rila Nat. Park
Christianity

Altindere Vadisi Nat. Park
Christianity

BULGARIA

GREECE

TURKEY

Agri Dagi Nat. Park
Christianity, Judaism

Göreme Nat. Park
Christianity

Meteora Monasteries
Christianity

see inset below

One of the monasteries of Meteora, Greece

Mt Kailash
Buddhism, Bön, Jainism, Hindusim

Khirganga, Mantalai Rakti Sar
Hinduism

NEPAL

Masang Khang
Bön, Buddhism

BHUTAN

Annapurna conservation area
Hindusim, Buddhism

Sagarmatha Nat. Park
Buddhism

CHINA

Mt Tai Shan
Daoism, Buddhism

Xishuangbanna
Buddhism

Phou Hin Poun
Traditional

LAOS

Chubu Sagaku Nat. Pk
Shinto

Nikko Nat. Park
Shinto, Buddhism

JAPAN

Mt Hakku San
Shinto, Buddhism

Sacred Forest of Kashima
Shinto

Kii Mt range
Shinto, Buddhism

Itsukushima
Shinto, Buddhism

Kinabulu mountain, Borneo

Socotra Island
Traditional

YEMEN

INDIA

Periyar Tiger Reserve
Hinduism

Sri Pada
Islam, Buddhism, Hinduism, Christianity

Phnom Kulen
Traditional

CAMBODIA

Angkor
Hinduism, Buddhism

Phnom Prich
Traditional

Phnom Nam Lyr
Traditional

Mihintale
Buddhism

SRI LANKA

Yala Nat. Park
Buddhism, Hinduism

Kataragama Sanctuary
Buddhism, Hinduism, Christianity, Islam

Sinharaja
Buddhism, Hinduism

PHILIPPINES

Mt Apo Nat. Park
Traditional

Kinabalu Nat. Park
Traditional

MALAYSIA

Tetepare
Traditional

SOLOMON ISLANDS

Hunstein Range
Traditional

Poboya-Palu Forest Park
Traditional

INDONESIA

PAPUA NEW GUINEA

Madang Lagoon
Traditional

Gunung Mutis
Traditional

Kakadu
Traditional

Mt Nyiro
Traditional

UGANDA

KENYA

Mt Kenya Nat. Park
Traditional, Christian

Kibale Nat. Park
Traditional

TANZANIA

Misali Island
Islam

MALAWI

Nyika Nat. Park
Traditional

Amber Mt. Nat. Park
Traditional

Tsodilo Hills
Traditional

MADAGASCAR

Sakoantovo
Traditional

BOTSWANA

Limpopo's Modjadji Reserve
Traditional

Lac Tsimanampetsotsa
Traditional

SOUTH AFRICA

Natal Drakensberg Nat. Park
Traditional

Mapungubwe Hill
Traditional

Harissa Forest
Christianity

Hermon River Nat. Res.
Christianity

LEBANON

Ouadi Qadisha, Forest of the Cedars of God
Christianity

Mt Carmel
Judaism

Baram
Judaism, Christianity

ISRAEL

EGYPT

St Catherine
Christianity, Islam, Judaism

St Catherine's Monastery area, Sinai

AUSTRALIA

Kata Tjuta
Traditional

Gulaga
Traditional

Deen Maar
Traditional

Tongariro Nat. Park
Traditional

NEW ZEALAND

	THE NATURE OF GOD	THE CREATION
BUDDHISM 	Some Buddhist recognize the existence of 'supernatural' or god-like beings, but Buddhists do not believe in an omnipotent creator God. All Buddhists recognize a transcendent truth and some conceive this in terms of a 'Buddha Nature' which infuses everything.	For Buddhists creation is cyclical, having no start and no end. It is part of the wheel of suffering to which we are attached through re-birth. Creation is seen as just part of this wheel.
CHRISTIANITY 	Christians believe in one God, creator of all things, considered to be three 'persons', the Trinity: God the Father, the Son (Jesus Christ) and the Holy Spirit. These three aspects of God co-exist within one Godhead.	All that exists does so through God, who began creation at a definite point in time and who will end creation. God created from nothing and all that he creates has purpose and meaning.
DAOISM 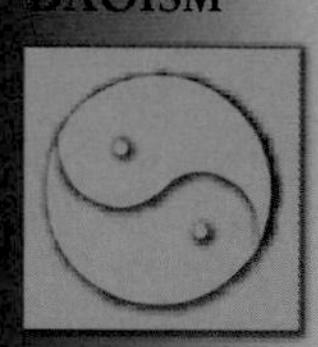	Daoists believe in universal forces of nature, known as Dao and expressed in two main forms: yin and yang. Through creative tension with each other they keep the world spinning and moving. Popular Daoism has thousands of gods, but Daoists do not believe in a single supreme God.	Creation as an event is not of great importance. There are various stories. In essence, the twin forces of yin and yang were created from nothing rather than by any being, and from these twin forces come all life.
HINDUISM 	Hindus believe in one Godhead or Divine Power, with innumerable forms. Three major forms are: Brahma, creator of each universe; Vishnu, sustainer and defender; and Shiva, destroyer and re-creator. Vishnu has 10 main forms, or avatars, which come to the help of the universe. These include Krishna and Rama.	Creation is cyclical. From the destruction of a previous universe, Brahma arises to create a new universe; Vishnu sustains it through a cycle of birth, growth and decline; Shiva destroys the universe and the cycle begins again.
ISLAM 	Muslims believe there is one God, Allah (Arabic for God). Allah is indivisible, has no equals, is the creator of all and has spoken to humanity through many prophets, of whom Muhammad is the last.	God is the creator. He simply says 'Be' and all things exist. God guides his creation and has a purpose for all forms of life within creation.
JUDAISM 	Jews believe in one God (whose name must not be pronounced), who created all things and who through his special covenant with the Jews has guided human life and destiny.	God is the creator, and the Book of Genesis says he created in six days and rested on the seventh. God will end creation in his own time.
SIKHISM 	Sikhs recognize one God, who is the true Guru (teacher). Unbound by time or space and beyond human definition, he makes himself known to those who are ready.	God is the creator of all, so all life is good. Attachment to this world means rebirth, so that release from this world is the highest goal.

Fundamentals of the Faiths

TIME	LIFE AFTER DEATH
Time is cyclical. Each existence continues through death and rebirth so long as the sense of self keeps us attached to this world. Individual desires are finally quenched (nirvana), but the world continues on its cyclical pattern. Some forms of Buddhism believe in a future Buddha who will come and bring release to all beings.	At death, each life continues in some other form – human, divine or animal, depending upon the results of behaviour in the last life. The goal of Buddhism is to extinguish the flame of wanting or attachment to the sense of self so that rebirth does not occur and nirvana is attained.
Time is linear, though there are two different approaches. One suggests that through human lives a renewed and peaceful world will be created: the Kingdom of God on earth. In the second, the world becomes so full of suffering and wrongdoing that an antichrist appears, bringing conflict. Christ returns and defeats the antichrist in a great battle, inaugurating a reign of peace.	There is one life only. Beliefs about death vary. The soul may ascend to heaven and be judged by God; or, the soul and the body may be raised on the Day of Judgement, at the end of time, and will then be judged.
There are elements of both the linear and the cyclical. There is no end to the world, just a personal journey, either to better and better rebirths, or into immortality.	At death, the soul is judged by up to 10 different gods of Hell, is purified by punishment then reborn again. Certain schools believe death is avoidable. By practising special meditations or eating certain things, the body is made immortal so that the person lives forever.
Time is cyclical. The world passes through various stages, from birth to growth to decline. We are currently in Kali Yuga, the age of decline. The world will eventually be destroyed, only for a new world to appear in the distant future.	Depending upon the karma – the consequences of action in this present life – at death, the soul (*atman*) is reborn in either a higher or lower physical form. Through devotion or correct behaviour it is possible to ascend through the orders of reincarnation, achieve liberation from the cycle of rebirth, and be reunited with the Supreme Being.
Time is linear. At the end of time, God will announce the Judgement Day and the world will end. All will be judged on that day.	There is one life only. After death, the individual awaits the Day of Judgement, when all will be brought back to life and judged. Paradise awaits those who have lived according to the will of God and those who have failed to do so cannot enter Paradise.
Time is linear. The Messiah, or the Chosen One of God, will come when either the world has become a better place or when it has reached the point of greatest trouble. The Messiah will herald an era of world peace.	There is one life only. Most religious Jews believe the individual awaits the Day of Judgement, when God will raise all to life and judgement. Some, however, believe that the soul is judged immediately after death.
Time is cyclical and beliefs associated with time are similar to those of Hinduism.	Each individual has many reincarnations, but being born a human means the soul is nearing the end of rebirth. God judges each soul at death and may either reincarnate the soul or, if pure enough, allow it to rest with him.

	SACRED LITERATURE	FOUNDERS AND PROPHETS
BUDDHISM 	The discourses of the Buddha, the rules for monks and nuns, and further knowledge are handed down in the 'Three Baskets' (*Tri-Pitaka*). Three versions survive: one in the Pali language (used by southern Buddhists) and two Mahayana versions in Chinese and Tibetan (used by northern Buddhists), which include later books not recognized as authoritative by southern Buddhists.	The Buddha was an Indian Prince, Siddhartha Gautama, who lived in the 6th century BCE. He was given the title Buddha, 'Awakened' or 'Enlightened One' when he understood the cause of suffering and the way to end suffering.
CHRISTIANITY 	The Bible consists of the Old Testament – the books of the Hebrew Bible – plus the New Testament. The books of the New Testament were fixed c 280 and are the Gospels (accounts of Jesus' life), early church history called the Acts of the Apostles, letters from early leaders such as Paul and James, and the Book of Revelation – a vision of the End of Time.	The faith is named after Jesus – called 'Christ', the Greek word for Chosen One – who was crucified c. 29 CE. Christians believe he is the Son of God, part of the Trinity and that he came to earth in human form to bring humanity back to fellowship with God.
DAOISM 	There are over 4,000 books in the Daoist Canon, from the 5th century BCE to the 14th century CE. Each school has its own favourites and many look back to the *Dao De Jing* of Lao Zi, compiled c. 5th century BCE, as their initial source of inspiration.	There have been various figures, ranging from mythical emperors to semi-historical figures such as Dao Zi (5th century BCE) and Zhang Dao Ling (2nd century CE), who founded popular Daoism.
HINDUISM 	There are many sacred books, including the teachings and history of the *Upanishads* and the *Puranas*, and the great epics of the *Ramayana* and *Mahabharata*. One of the most famous and popular religious writings, the *Bhagavad Gita* is contained in the *Mahabharata*.	There are thousands of Hindu gurus, reflecting the huge variety of teachings. A guru, or teacher, is someone who has gained wisdom and clarity through knowledge and practice. A Hindu wanting to follow a particular path of prayer, meditation and devotion usually has a guru.
ISLAM 	The Qur'an, dictated by the Angel Jibra'il to Muhammad in the first part of the 7th century CE. Muslims believe the Qur'an was written by God before time began.	Islam means to be in submission to God. There have been numerous prophets who came to remind people of God's will, such as Abraham, Moses and Jesus. The final prophet is believed to be Muhammad, who lived in the 6th–7th century CE.
JUDAISM 	The Hebrew Bible has three parts: the Torah (Five Books of Moses), the Prophets and the Writings such as Esther and the Psalms. The Torah contains laws, doctrine and guidance on way of life, as well as accounts of the early history of the Israelites and their relationship with God.	Through the covenant with Abraham and his descendants, God selected the Israelites as his chosen people. This covenant was reaffirmed and consolidated with Moses, when God gave Moses the Law by which Israelites were to live.
SIKHISM 	The Guru Granth Sahib, a collection of writings and hymns by some of the 10 Gurus of Sikhism, plus material from Muslim and Hindu writers, was compiled mid-16th century. It was made the eleventh and final Guru of Sikhism at the death of the tenth Guru in 1708.	Guru Nanak (1469–1539) was the first Guru of Sikhism and was followed by nine more human Gurus. The tenth and last was Guru Gobind Singh (1675–1780), who appointed the Scriptures, the Granth Sahib, as the final Guru.

Fundamentals of the Faiths

RITES OF BIRTH AND DEATH	FESTIVALS
Buddhists invite monks and nuns to attend such events and to read scriptures, but the main ceremonial rituals are often linked to other cultural or religious traditions. In Theravada Buddhism, funerals are occasions for teaching about suffering and impermanence and for chanting *paritta* (protection) in order to gain and transfer merit for the sake of the deceased.	Buddhist festivals follow the lunar cycle and include Vesakha Puja, which celebrates the birth, enlightenment and passing into nirvana of the Buddha (May–June). Alsalha Puja celebrates the Buddha's first sermon, when he taught the principles of Buddhism (July–Aug), and the lunar New Year is a time to sweep away the negative aspects of the past year. 
Many Christians are baptized into the Church while they are babies, but this can be done at any time in life. At death, Christians are laid to rest in the hope of the resurrection of the dead. Cremation or burial are both acceptable.	The main festivals commemorate the life of Jesus Christ: Christmas, celebrating his birth (25 Dec); Easter, marking his death and resurrection (March–April); Ascension Day, celebrating his return to Heaven (May). Pentecost celebrates the coming of the Holy Spirit onto the Disciples (May–June). 
Horoscopes are cast at birth. After a month a naming ceremony is held. At death, the body is buried and paper models of money, houses and cars are burnt to help the soul in the afterlife. After about 10 years the body is dug up and the bones buried again in an auspicious site.	There are hundreds of local festivals. The main festivals: Chinese New Year (Jan–Feb); Qing Ming, for the veneration of the dead (4 or 5 April); the Hungry Ghosts' festival for the release of the restless dead; the Moon Festival, celebrating the harvest moon (Sept–Oct). 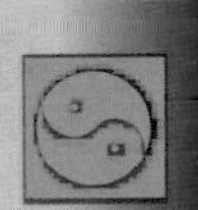
Before birth and in the first months of life, there are many ceremonies. These include: reciting the scriptures to the baby in the womb; casting its horoscope when it is born; cutting its hair for the first time. At death, bodies are cremated and the ashes thrown on to a sacred river. The River Ganges is the most sacred river of all.	There are many festivals. Major ones include: Mahashivaratri, celebrating Shiva (Feb); Holi, the harvest festival in honour of love and of Krishna (Feb–March); Divali, the festival of lights celebrating New Year, honouring Lakshmi, the goddess of wealth, and recalling the triumphant return of Rama and Sita from exile, as told in the epic *Ramayana* (Oct–Nov). 
At birth, the call to prayer is whispered into the baby's ear. After seven days the baby is given a name, shaved, and baby boys are circumcized. At death, the body is washed as if ready for prayer and then buried as soon as possible. Cremation is not allowed.	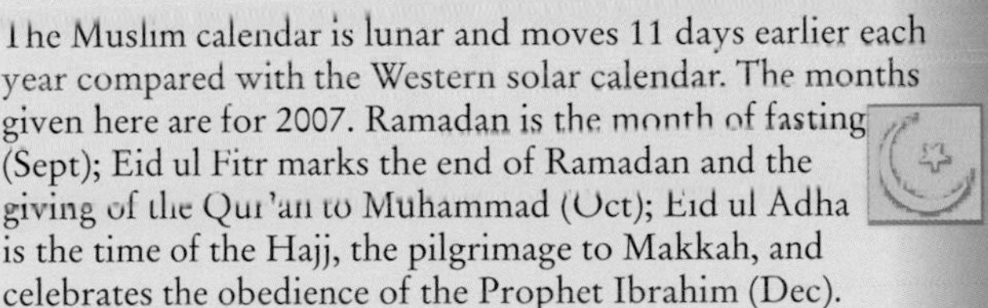 The Muslim calendar is lunar and moves 11 days earlier each year compared with the Western solar calendar. The months given here are for 2007. Ramadan is the month of fasting (Sept); Eid ul Fitr marks the end of Ramadan and the giving of the Qur'an to Muhammad (Oct); Eid ul Adha is the time of the Hajj, the pilgrimage to Makkah, and celebrates the obedience of the Prophet Ibrahim (Dec).
Baby boys are circumcized eight days after birth. The names of girls are announced in the synagogue on the first Sabbath after birth. Burial takes place within 24 hours of death and cremation is very rare. The family is in full mourning for seven days and, for eleven months, the special prayer Kadish is said every day.	Passover or Pesach celebrates the Exodus of the Jews from Egypt (March–April); Shavuot, or Pentecost, marks the giving of the Law to Moses (May–June); Rosh Hashanah is the New Year festival, and Yom Kippur, the day of repentance (both in Sept–Oct); Hanukah celebrates the survival of the Jews (Dec). 
At birth, the Mool mantra, the core teaching of Sikhism, is whispered into the baby's ear. The baby is named at the gurdwara, or place of worship. The Guru Granth Sahib is opened and the first letter of the first word on the page gives the first letter of the baby's name. At death, the body is cremated and the ashes thrown on to running water.	Baisakhi celebrates the foundation of the Khalsa (13 April). Other major festivals include the Martyrdom of Guru Arjan Dev (May–June); the birthday of Guru Nanak, the founder of Sikhism (Oct-Nov); the Martyrdom of Guru Tegh Bahadur (Nov–Dec); and the birthday of Guru Gobind Singh (Dec-Jan). 

Part Seven SOCIAL CONTEXT

The vast majority of the world's population belong to a religion, and therefore experience ritual, mystery, music, art, drama, belief and sacred places. Most people also live in what is sometimes rather patronisingly called 'the real world' of economics, politics, class, race and other such social issues. The graphics in this book give you a glimpse of the religious world and the way it is organized, but it is helpful to place this within a wider social context.

Huge social changes are taking place: half of all people now live in cities, and access to mass media is changing the way we understand ourselves and those around us. The internet allows access to information that would have been unimaginable just a few years ago, and no government can now control the flow of ideas. Within all this, the role and significance of the religions is a story not often heard, but charted and tracked in some detail in this book. Together with the wider social data contained in this section, it is possible to begin to see how the world of today and tomorrow is taking shape, and where the faiths may be able to continue their millennia-long role of helping to make sense of it and define our role within it.

A Tibetan Buddhist using prayer beads during meditation at a UK shrine

Countries	Population			Surface area	Gross National Income (GNI)	Life expectancy
	thousands *2005 or latest available data*	annual % growth *2005*	urban as % of total *2005*	square kilometres *2005*	$ per capita *2005*	at birth *2004*
Afghanistan	–	–	–	652,090	–	–
Albania	3,130	0.58%	44%	28,750	2,580	74
Algeria	32,854	1.52%	59%	2,381,740	2,730	71
Angola	15,941	2.87%	36%	1,246,700	1,350	41
Antigua and Barbuda	81	1.14%	38%	440	10,920	–
Argentina	38,747	0.97%	90%	2,780,400	4,470	75
Armenia	3,016	-0.32%	65%	29,800	1,470	71
Australia	20,321	1.04%	92%	7,741,220	32,220	80
Austria	8,211	0.47%	66%	83,870	36,980	79
Azerbaijan	8,388	0.98%	50%	86,600	1,240	72
Bahrain	727	1.50%	90%	710	–	75
Bangladesh	141,822	1.86%	24%	144,000	470	63
Barbados	270	0.25%	52%	430	–	75
Belarus	9,776	-0.50%	71%	207,600	2,760	68
Belgium	10,471	0.47%	97%	32,545	35,700	79
Belize	292	3.20%	48%	22,970	3,500	72
Benin	8,439	3.15%	45%	112,620	510	55
Bhutan	918	2.42%	9%	47,000	870	64
Bolivia	9,182	1.90%	63%	1,098,580	1,010	65
Bosnia and Herzegovina	3,907	-0.06%	44%	51,210	2,440	74
Botswana	1,765	-0.23%	52%	581,730	5,180	35
Brazil	186,405	1.35%	83%	8,514,880	3,460	71
Brunei	374	2.20%	76%	5,770	–	77
Bulgaria	7,741	-0.26%	70%	110,990	3,450	72
Burkina Faso	13,228	3.12%	18%	274,000	400	48
Burma	50,519	1.03%	30%	676,580	–	61
Burundi	7,548	3.58%	10%	27,830	100	44
Cambodia	14,071	1.96%	19%	181,040	380	57
Cameroon	16,322	1.76%	51%	475,440	1,010	46
Canada	32,271	0.92%	80%	9,984,670	32,600	80
Cape Verde	507	2.32%	56%	4,030	1,870	70
Central African Republic	4,038	1.29%	43%	622,980	350	39
Chad	9,749	3.14%	25%	1,284,000	400	44
Chile	16,295	1.06%	87%	756,630	5,870	78
China	1,304,500	0.64%	39%	9,598,060	1,740	71
Colombia	45,600	1.51%	76%	1,138,910	2,290	73
Comoros	600	2.11%	35%	2,230	640	63
Congo	3,999	2.94%	54%	342,000	950	52
Congo, Dem. Rep.	57,549	2.99%	32%	2,344,860	120	44
Cook Islands	21	-1.20%	–	237	9,100	–
Costa Rica	4,327	1.73%	61%	51,100	4,590	79
Côte d'Ivoire	18,154	1.57%	45%	322,460	840	46
Croatia	4,444	0.05%	59%	56,540	8,060	75
Cuba	11,269	0.22%	76%	110,860	–	77
Cyprus	835	1.13%	69%	9,250	–	79
Czech Republic	10,196	-0.19%	74%	78,870	10,710	76
Denmark	5,418	0.26%	85%	43,090	47,390	77
Djibouti	793	1.78%	84%	23,200	1,020	53

Mortality rate deaths of under-5s per 1,000 *2004*	**HIV prevalence** % of population ages 15–49 *2005*	**Adult literacy** % of people ages 15 and above *2004*	**Poverty** % of population below national poverty line *2002 or latest available*	**Televisions** number per 1,000 people *2000*	**Internet** users per 1,000 people *2004*	**Countries**
–	0.10%	28%	–	14	1	Afghanistan
19	–	99%	–	123	24	Albania
40	0.09%	70%	12%	110	26	Algeria
260	3.68%	67%	–	19	11	Angola
12	–	–	–	–	250	Antigua and Barbuda
18	0.61%	97%	–	293	133	Argentina
32	0.15%	99%	–	244	50	Armenia
6	0.10%	–	–	738	646	Australia
5	0.29%	–	–	536	477	Austria
90	0.11%	99%	–	259	49	Azerbaijan
11	–	87%	–	–	213	Bahrain
77	0.10%	–	50%	7	2	Bangladesh
12	1.55%	–	–	–	558	Barbados
11	0.34%	100%	0%	342	163	Belarus
5	0.27%	–	0%	541	403	Belgium
39	2.49%	–	–	–	124	Belize
152	1.79%	35%	33%	45	12	Benin
80	0.10%	–	–	–	22	Bhutan
69	0.13%	87%	63%	119	39	Bolivia
15	0.10%	97%	–	111	58	Bosnia and Herzegovina
116	24.10%	81%	–	25	34	Botswana
34	0.54%	89%	17%	343	120	Brazil
9	0.10%	93%	–	–	153	Brunei
15	0.10%	98%	–	449	283	Bulgaria
192	2.01%	22%	45%	12	4	Burkina Faso
106	1.25%	90%	–	7	1	Burma
190	3.26%	59%	–	30	3	Burundi
141	1.64%	74%	36%	8	3	Cambodia
149	5.43%	68%	40%	34	10	Cameroon
6	0.30%	–	–	–	626	Canada
36	–	–	–	–	50	Cape Verde
193	10.73%	49%	–	6	2	Central African Republic
200	3.52%	26%	64%	1	6	Chad
8	0.28%	96%	17%	242	267	Chile
31	0.08%	91%	5%	293	73	China
21	0.61%	93%	64%	282	80	Colombia
70	0.10%	–	–	–	14	Comoros
108	5.27%	–	–	13	9	Congo
205	3.23%	67%	–	2	–	Congo, Dem. Rep.
–	–	95%	–	–	6	Cook Islands
13	0.29%	95%	22%	231	235	Costa Rica
194	7.06%	49%	37%	60	17	Côte d'Ivoire
7	0.10%	98%	–	293	293	Croatia
7	0.09%	100%	–	250	13	Cuba
5	–	97%	–	–	361	Cyprus
4	0.10%	–	–	508	470	Czech Republic
5	0.22%	–	–	807	696	Denmark
126	3.11%	–	45%	–	12	Djibouti

Countries	Population thousands *2005 or latest available data*	Population annual % growth *2005*	Population urban as % of total *2005*	Surface area square kilometres *2005*	Gross National Income (GNI) $ per capita *2005*	Life expectancy at birth *2004*
Dominica	72	0.75%	72%	750	3,790	–
Dominican Republic	8,895	1.44%	59%	48,730	2,370	68
East Timor	976	5.36%	8%	14,870	750	–
Ecuador	13,228	1.43%	62%	283,560	2,630	75
Egypt	74,033	1.90%	42%	1,001,450	1,250	70
El Salvador	6,881	1.74%	59%	21,040	2,450	71
Equatorial Guinea	504	2.27%	48%	28,050	–	43
Eritrea	4,401	3.93%	20%	117,600	220	54
Estonia	1,345	-0.30%	70%	45,230	9,100	72
Ethiopia	71,256	1.83%	16%	1,104,300	160	42
Fiji	848	0.82%	52%	18,270	3,280	68
Finland	5,245	0.32%	61%	338,150	37,460	79
France	60,743	0.60%	76%	551,500	34,810	80
Gabon	1,384	1.57%	84%	267,670	5,010	54
Gambia	1,517	2.63%	26%	11,300	290	56
Georgia	4,474	-0.97%	52%	69,700	1,350	71
Germany	82,485	-0.04%	88%	357,030	34,580	78
Ghana	22,113	2.05%	45%	238,540	450	57
Greece	11,089	0.29%	61%	131,960	19,670	79
Grenada	107	0.71%	41%	340	3,920	–
Guam	170	1.70%	–	550	15,000	75
Guatemala	12,599	2.44%	46%	108,890	2,400	68
Guinea	9,402	2.15%	35%	245,860	370	54
Guinea-Bissau	1,586	2.98%	34%	36,120	180	45
Guyana	751	0.13%	38%	214,970	1,010	64
Haiti	8,528	1.43%	38%	27,750	450	52
Honduras	7,205	2.19%	46%	112,090	1,190	68
Hungary	10,088	-0.19%	65%	93,030	10,030	73
Iceland	295	1.03%	93%	103,000	46,320	80
India	1,094,583	1.37%	28%	3,287,260	720	63
Indonesia	220,558	1.36%	46%	1,904,570	1,280	67
Iran	67,700	1.03%	67%	1,648,200	2,770	71
Iraq	–	–	–	438,320	–	–
Ireland	4,151	2.00%	60%	70,270	40,150	78
Israel	6,909	1.62%	92%	22,140	18,620	79
Italy	57,471	-0.18%	67%	301,340	30,010	80
Jamaica	2,657	0.48%	52%	10,990	3,400	71
Japan	127,956	0.15%	66%	377,900	38,980	82
Jordan	5,411	2.56%	79%	88,780	2,500	72
Kazakhstan	15,146	0.88%	56%	2,724,900	2,930	65
Kenya	34,256	2.33%	39%	580,370	530	48
Kiribati	99	1.21%	–	730	1,390	–
Korea (North)	22,488	0.46%	–	120,540	–	64
Korea (South)	48,294	0.44%	80%	99,260	15,830	77
Kuwait	2,535	3.04%	96%	17,820	–	77
Kyrgyzstan	5,156	1.23%	34%	199,900	440	68
Laos	5,924	2.26%	21%	236,800	440	55
Latvia	2,300	-0.55%	66%	64,590	6,760	71

Mortality rate deaths of under-5s per 1,000 *2004*	HIV prevalence % of population ages 15–49 *2005*	Adult literacy % of people ages 15 and above *2004*	Poverty % of population below national poverty line *2002 or latest available*	Televisions number per 1,000 people *2000*	Internet users per 1,000 people *2004*	Countries
14	–	–	–	–	259	Dominica
32	1.11%	87%	29%	97	91	Dominican Republic
80	–	–	–	–	–	East Timor
26	0.29%	91%	35%	218	48	Ecuador
36	0.10%	71%	17%	189	54	Egypt
28	0.92%	–	48%	201	87	El Salvador
204	3.20%	87%	–	–	10	Equatorial Guinea
82	2.36%	–	53%	26	12	Eritrea
8	1.32%	100%	–	591	497	Estonia
166	–	–	44%	6	2	Ethiopia
20	0.11%	–	–	–	73	Fiji
4	0.08%	–	–	692	629	Finland
5	0.40%	–	–	620	414	France
91	7.88%	–	–	326	29	Gabon
122	2.44%	–	64%	3	33	Gambia
45	0.22%	–	–	474	39	Georgia
5	0.12%	–	–	586	500	Germany
112	2.27%	58%	40%	118	17	Ghana
5	0.17%	96%	–	488	177	Greece
21	–	–	–	–	76	Grenada
–	–	99%	23%	–	474	Guam
45	0.90%	69%	56%	61	61	Guatemala
155	1.52%	29%	40%	44	5	Guinea
203	3.79%	–	49%	–	17	Guinea-Bissau
64	2.45%	–	35%	–	193	Guyana
117	3.81%	–	65%	5	59	Haiti
41	1.54%	80%	53%	96	32	Honduras
8	0.06%	–	–	437	267	Hungary
3	0.20%	–	–	–	772	Iceland
85	0.92%	61%	29%	78	32	India
38	0.13%	90%	27%	149	67	Indonesia
38	0.15%	77%	–	163	82	Iran
–	–	74%	–	83	1	Iraq
6	0.23%	–	–	399	265	Ireland
6	–	97%	–	335	471	Israel
5	0.50%	98%	–	494	501	Italy
20	1.53%	80%	19%	194	403	Jamaica
4	0.10%	–	–	725	587	Japan
27	–	90%	12%	84	114	Jordan
73	0.10%	100%	–	241	27	Kazakhstan
120	6.09%	74%	42%	25	45	Kenya
65	–	–	–	–	20	Kiribati
55	–	–	–	54	–	Korea (North)
6	0.10%	–	–	364	657	Korea (South)
12	–	93%	–	486	244	Kuwait
68	0.14%	99%	–	49	52	Kyrgyzstan
83	0.12%	69%	39%	10	4	Laos
12	0.79%	100%	–	789	350	Latvia

Countries	Population thousands *2005 or latest available data*	annual % growth *2005*	urban as % of total *2005*	Surface area square kilometres *2005*	Gross National Income (GNI) $ per capita *2005*	Life expectancy at birth *2004*
Lebanon	3,577	1.03%	88%	10,400	6,180	72
Lesotho	1,795	-0.18%	18%	30,350	960	36
Liberia	3,283	1.31%	–	111,370	130	42
Libya	5,853	1.95%	86%	1,759,540	5,530	74
Lithuania	3,415	-0.60%	67%	65,300	7,050	72
Luxembourg	457	0.75%	92%	–	65,630	78
Macedonia	2,034	0.18%	60%	25,710	2,830	74
Madagascar	18,606	2.69%	27%	587,040	290	56
Malawi	12,884	2.16%	16%	118,480	160	40
Malaysia	25,347	1.80%	64%	329,740	4,960	73
Maldives	329	2.46%	29%	300	2,390	67
Mali	13,518	2.96%	32%	1,240,190	380	48
Malta	404	0.70%	92%	320	13,590	79
Marshall Islands	63	3.29%	–	180	2,930	–
Mauritania	3,069	2.92%	62%	1,025,520	560	53
Mauritius	1,248	1.11%	43%	2,040	5,260	73
Mexico	103,089	1.01%	76%	1,958,200	7,310	75
Micronesia, Fed. Sts.	110	0.72%	–	700	2,300	68
Moldova	4,206	-0.29%	46%	33,840	880	68
Mongolia	2,554	1.55%	57%	1,566,500	690	65
Morocco	30,168	1.15%	57%	446,550	1,730	70
Mozambique	19,792	1.88%	36%	801,590	310	42
Namibia	2,031	1.09%	32%	824,290	2,990	47
Nepal	27,133	2.02%	15%	147,180	270	62
Netherlands	16,329	0.29%	66%	41,530	36,620	79
New Caledonia	234	1.90%	–	18,580	–	75
New Zealand	4,110	1.20%	86%	270,530	25,960	79
Nicaragua	5,487	2.04%	57%	130,000	910	70
Niger	13,957	3.34%	22%	1,267,000	240	45
Nigeria	131,530	2.17%	47%	923,770	560	44
Norway	4,618	0.59%	79%	323,760	59,590	80
Oman	2,567	1.30%	78%	309,500	–	75
Pakistan	155,772	2.41%	34%	796,100	690	65
Palestine Authority	3,626	3.31%	71%	–	–	73
Palau	20	1.00%	–	–	7,630	–
Panama	3,232	1.75%	57%	75,520	4,630	75
Papua New Guinea	5,887	1.98%	13%	462,840	660	56
Paraguay	6,158	2.32%	57%	406,750	1,280	71
Peru	27,968	1.46%	74%	1,285,220	2,610	70
Philippines	83,054	1.75%	61%	300,000	1,300	71
Poland	38,165	-0.04%	62%	312,690	7,110	74
Portugal	10,557	0.52%	55%	91,980	16,170	77
Puerto Rico	3,911	0.42%	–	8,950	–	77
Qatar	813	4.52%	92%	11,000	–	74
Romania	21,632	-0.24%	55%	238,390	3,830	71
Russia	143,151	-0.49%	73%	17,098,240	4,460	65
Rwanda	9,038	1.73%	19%	26,340	230	44
Samoa	185	0.67%	22%	2,840	2,090	70

Mortality rate deaths of under-5s per 1,000 *2004*	HIV prevalence % of population ages 15–49 *2005*	Adult literacy % of people ages 15 and above *2004*	Poverty % of population below national poverty line *2002 or latest available*	Televisions number per 1,000 people *2000*	Internet users per 1,000 people *2004*	Countries
31	0.14%	–	–	335	169	Lebanon
112	23.24%	82%	49%	16	24	Lesotho
235	–	–	–	25	–	Liberia
20	–	–	–	137	36	Libya
8	0.17%	100%	–	422	282	Lithuania
6	0.20%	–	–	–	597	Luxembourg
14	0.10%	96%	–	282	78	Macedonia
123	0.51%	71%	71%	24	5	Madagascar
175	14.09%	64%	65%	3	4	Malawi
12	0.47%	89%	16%	168	397	Malaysia
46	–	96%	–	–	59	Maldives
219	1.73%	19%	64%	14	4	Mali
6	0.13%	88%	–	–	750	Malta
59	–	–	–	–	33	Marshall Islands
125	0.68%	51%	40%	96	6	Mauritania
15	0.55%	84%	11%	268	146	Mauritius
28	0.28%	91%	10%	283	138	Mexico
23	–	–	–	–	109	Micronesia, Fed. Sts.
28	1.05%	98%	–	297	96	Moldova
52	0.10%	98%	36%	65	80	Mongolia
43	0.10%	52%	19%	166	117	Morocco
152	16.11%	–	69%	5	7	Mozambique
63	19.56%	85%	–	38	37	Namibia
76	0.53%	49%	42%	7	7	Nepal
6	0.22%	–	–	538	614	Netherlands
–	–	96%	–	–	304	New Caledonia
7	0.10%	–	–	522	788	New Zealand
38	0.24%	77%	48%	69	23	Nicaragua
259	1.10%	29%	63%	37	2	Niger
197	3.86%	–	34%	68	14	Nigeria
4	0.11%	–	–	669	390	Norway
13	–	81%	–	563	97	Oman
101	0.10%	50%	33%	131	13	Pakistan
–	–	92%	–	–	46	Palestine Authority
27	–	–	–	–	–	Palau
24	0.89%	92%	37%	194	94	Panama
93	1.76%	57%	38%	17	29	Papua New Guinea
24	0.38%	–	22%	218	25	Paraguay
29	0.57%	88%	49%	148	117	Peru
34	0.10%	93%	37%	144	54	Philippines
8	0.12%	–	–	400	236	Poland
5	0.40%	–	–	630	281	Portugal
–	–	–	–	–	221	Puerto Rico
21	–	89%	–	–	212	Qatar
20	0.10%	97%	–	381	208	Romania
21	1.09%	99%	–	421	111	Russia
203	3.07%	65%	51%	0	4	Rwanda
30	–	–	–	–	33	Samoa

Countries	Population			Surface area	Gross National Income (GNI)	Life expectancy
	thousands *2005 or latest available data*	annual % growth *2005*	urban as % of total *2005*	square kilometres *2005*	$ per capita *2005*	at birth *2004*
São Tomé and Principe	157	2.30%	38%	960	390	63
Saudi Arabia	24,573	2.57%	88%	2,149,690	11,770	72
Senegal	11,658	2.36%	50%	196,720	710	56
Serbia and Montenegro	8,168	0.26%	–	102,170	3,280	73
Seychelles	84	1.01%	50%	460	8,290	–
Sierra Leone	5,525	3.48%	39%	71,740	220	41
Singapore	4,351	2.59%	100%	680	27,490	79
Slovakia	5,387	0.09%	58%	49,030	7,950	74
Slovenia	1,998	0.06%	51%	20,270	17,350	77
Solomon Islands	478	2.53%	17%	28,900	590	63
Somalia	8,228	3.25%	–	637,660	–	47
South Africa	45,192	-0.70%	57%	1,219,090	4,960	45
Spain	43,389	1.62%	77%	505,370	25,360	80
Sri Lanka	19,582	0.84%	21%	65,610	1,160	74
St. Kitts and Nevis	48	2.14%	32%	360	8,210	–
St. Lucia	166	1.12%	31%	620	4,800	73
St. Vincent and the Grenadines	119	0.52%	58%	390	3,590	71
Sudan	36,233	1.98%	39%	2,505,810	640	57
Suriname	449	0.62%	76%	163,270	2,540	69
Swaziland	1,131	0.99%	24%	17,360	2,280	42
Sweden	9,024	0.36%	83%	450,290	41,060	81
Switzerland	7,441	0.69%	68%	41,280	54,930	81
Syria	19,043	2.45%	50%	185,180	1,380	74
Tajikistan	6,507	1.19%	25%	142,550	330	64
Tanzania	38,329	1.85%	35%	945,090	340	46
Thailand	64,233	0.84%	32%	513,120	2,750	71
Togo	6,145	2.58%	35%	56,790	350	55
Tonga	102	0.32%	34%	750	2,190	72
Trinidad and Tobago	1,305	0.30%	75%	5,130	10,440	70
Tunisia	10,022	0.90%	64%	163,610	2,890	73
Turkey	72,636	1.26%	66%	783,560	4,710	70
Turkmenistan	4,833	1.40%	45%	488,100	–	63
Uganda	28,816	3.52%	12%	241,040	280	49
Ukraine	47,111	-0.72%	67%	603,700	1,520	68
United Arab Emirates	4,533	4.82%	85%	83,600	–	79
United Kingdom	60,203	0.56%	89%	243,610	37,600	79
United States	296,497	0.96%	80%	9,629,090	43,740	77
Uruguay	3,463	0.69%	93%	176,220	4,360	75
Uzbekistan	26,593	1.45%	37%	447,400	510	67
Vanuatu	211	1.93%	23%	12,190	1,600	69
Venezuela	26,577	1.71%	88%	912,050	4,810	74
Vietnam	82,966	0.97%	26%	331,690	620	70
Virgin Islands (US)	115	1.63%	–	350	–	79
Yemen	20,975	3.12%	26%	527,970	600	61
Zambia	11,668	1.64%	36%	752,610	490	38
Zimbabwe	13,010	0.56%	35%	390,760	340	37

Sources: Urban population: *Human Development Report 2005* hdr.undp.org/reports/global/2005; Televisions: International Telecommunications Union; All other data: World Development Indicators www.worldbank.org/data. Data for Cook Islands, Palestine Authority, and selected data for Guam from *CIA Factbook*

Mortality rate deaths of under-5s per 1,000 *2004*	**HIV prevalence** % of population ages 15–49 *2005*	**Adult literacy** % of people ages 15 and above *2004*	**Poverty** % of population below national poverty line *2002 or latest available*	**Televisions** number per 1,000 people *2000*	**Internet** users per 1,000 people *2004*	**Countries**
118	–	–	–	–	131	São Tomé and Principe
27	–	79%	–	264	66	Saudi Arabia
137	0.91%	39%	33%	40	42	Senegal
15	0.19%	96%	–	–	147	Serbia and Montenegro
14	–	92%	–	–	239	Seychelles
283	1.56%	35%	68%	13	2	Sierra Leone
3	0.30%	93%	–	304	571	Singapore
9	0.10%	100%	–	407	423	Slovakia
4	0.10%	–	–	368	476	Slovenia
56	–	–	–	–	6	Solomon Islands
225	0.86%	–	–	14	25	Somalia
67	18.78%	82%	–	127	78	South Africa
5	0.62%	–		591	336	Spain
14	0.10%	91%	25%	111	14	Sri Lanka
–	–	–	–	–		St. Kitts and Nevis
14	–	–	–	–	336	St. Lucia
22	–	–	–	–	68	St. Vincent and the Grenadines
91	1.59%	61%	–	273	32	Sudan
39	1.94%	90%	–	–	67	Suriname
156	33.38%	80%	40%	119	32	Swaziland
4	0.19%	–	–	574	756	Sweden
5	0.40%	–	–	548	474	Switzerland
16	–	80%	–	67	43	Syria
93	0.14%	99%	–	326	1	Tajikistan
126	6.46%	69%	36%	20	9	Tanzania
21	1.40%	93%	13%	284	109	Thailand
140	3.24%	53%	32%	32	37	Togo
25	–	99%	–	–	29	Tonga
20	2.64%	–	21%	340	123	Trinidad and Tobago
25	0.13%	74%	8%	198	84	Tunisia
32	–	87%	–	449	142	Turkey
103	0.10%	99%	–	196	8	Turkmenistan
138	6.66%	67%	55%	27	7	Uganda
18	1.40%	99%	–	456	79	Ukraine
8	–	–	–	292	321	United Arab Emirates
6	–	–	–	653	628	United Kingdom
8	0.60%	–	–	854	630	United States
17	0.49%	–	–	530	198	Uruguay
69	0.21%	–	–	276	34	Uzbekistan
40	–	74%	–	–	36	Vanuatu
19	0.72%	93%	31%	185	89	Venezuela
23	0.51%	90%	51%	185	71	Vietnam
–	–	–	–	–	–	Virgin Islands (US)
111	–	–	42%	283	9	Yemen
182	16.96%	68%	73%	134	20	Zambia
129	20.12%	–	35%	30	63	Zimbabwe

Notes and Sources

The sources cited here relate to the maps, graphics and all text.

Popular Religions

(pages 14–15)

Belonging to a religion is, for those such as Jews, Zoroastrians and in most cases Sikhs, a matter of ethnic identity. For others, belonging to a religion is an issue of personal choice – a factor that is of increasing significance as the world becomes more pluralist. For yet others, it is a matter of following local, family or cultural norms and is as ordinary as waking and sleeping; it simply is part of being a human being in a given context.

Religion is a highly complex series of commitments, habits, customs, choices, desires and intentions, and as such it affects some 80 percent of the world's population. For many, the idea that religion is a separate set of beliefs from the beliefs and practices used in everyday life is simply unimaginable. The notion that religion is private and should not affect social or even economic life makes no sense for most Muslims, for example, for whom religion means how to be a human being and how to live a good life.

In Christianity, the notion of the possibility, even desirability, of separating religion from secular life is largely a Protestant northern European/North American idea, but its impact has been immense. Because much academic study of religion has sprung from these two regions, the Protestant division into a public, secular world and a private, religious world has coloured how religion is understood. This has, in turn, led to a greater marginalization of religion, and not just in those areas where the secular–religious division is an historic one. Modern India has tried to restrain or contain religion by using this division. The secular constitution it adopted on independence in 1947 was a direct legacy of rule by a colonial power, and this artificial divide may well be contributing to present tensions in India.

Where religions are expanding fast, religious commitment often carries with it a powerful social, political and ethnic identity. A change in religion is often also a social and political statement. This can be seen most clearly in the states that replaced the former USSR, those experiencing a resurgence of Islam, and in the general upsurge of religion in Africa.

In Northern Europe, Australasia and Canada the situation is very different. While a majority claim allegiance to a religion, in countries such as Sweden or Norway fewer than 10 percent attend church on a regular basis. This secularization of behaviour is called 'European Exceptionalism', as the phenomenon is largely restricted to Europe and its 'satellites' of Canada and Australasia. It does not mean those not attending church do not feel themselves 'Christian', or entitled to the rituals and services of the religion. The recent growth of interest in religious and spiritual issues in 'secular' cultures is testimony to a growing willingness to explore such areas, albeit not necessarily through conventional channels. Nevertheless, there is in these countries a continued steady growth in secularism – also often to be found in the educated urban classes outside Western Europe. The USA, while to a great extent sharing what is seen as a common 'secular' culture with Northern Europe, has a high level of church attendance amongst those professing a religion. However it would be a mistake to equate church attendance with a religious life.

TV and radio are another common means of worship throughout Europe. Church attendance in Norway, for example, has historically never been high due to the distances between places of worship and extremes of climate. But over 50 percent listen to the Sunday morning service on radio or TV. In Western Europe overall, however, the decline in weekly attendance at a place of worship does indicate a decline of religious observance.

Sources

David B Barrett, George T Kurian, Todd M Johnson, *World Christian Encyclopedia*, vol 1, 2nd edition, Oxford University Press, 2001

David B Barrett, Todd M Johnson, *World Christian Trends AD30–AD2200*, William Carey Library, Pasadena, California, 2001

Chris Horrie and Peter Chippendale, *What is Islam, A Comprehensive Introduction*, Virgin Books, 2003

International Religious Freedom Reports 2006, US Department of State, www.state.gov/g/drl/rls/irf//2006/

Arrivals

(pages 16–17)

By 1450, Christianity appeared to be almost exclusively a white European religion, while Islam, with its heartlands in the Middle East, was primarily an Arab and Turkic religion and was the larger of the two in terms of geographical spread and probably also numerically. From the 1450s onwards, Christianity began its most dramatic spread since the time of the Roman Empire, and within 400 years was the world's largest religion – a position it holds to this day. Christianity initially began its expansion by sea, in part to avoid the Muslim empires that lay between Europe and the Far East, with its spices and riches. Advances in navigation and the voyages of Henry the Navigator from the mid-15th century onwards opened up new trade routes. Portuguese sailors crept their way round the northwest coast of Africa, making the first new European landfall in West Africa in 1445, explaining the early arrival of Christianity in Equatorial Guinea. These early voyages opened the way for the major expeditions of the late 15th century.

Following the capture of Constantinople in 1453, Islam spread more deeply into Europe. At one point in the 17th century, the Muslim Ottoman Empire controlled Greece, Bulgaria, Romania, Hungary and the area of former Yugoslavia. This proved to be a passing phase. More significant was its deeper penetration into parts of sub-Saharan Africa, and into South-East Asia, consolidating and expanding its presence in both areas throughout the last 400 years, often as a response to the increase in Christian missionary work that followed the explorers' ships.

The development of Christianity was the more dramatic. Using the newly discovered trade routes across the seas, Western European nations such as Spain, Portugal, England, France and the Netherlands, took both Christianity and colonialism around the world. The 'discovery' of the Americas was to have the greatest impact, but the spread of Christianity around the coastline of West, Southern and East Africa, and into parts of South-East Asia marked a major shift in the religious alignment of these large areas. By land, Russia was to colonize and Christianize Siberia from the 1600s onwards.

At that time, most of the rest of the world, including the Americas, Australasia and the majority of Africa, practised local traditional religions, largely unaffected by any major world religion. Unlike Christianity or Islam, none of these cultures made universal claims for their teachings, nor did they experience the missionary impetus that propelled Christianity and Islam across the world.

The expansion of Christianity and Islam dramatically altered the world. Christianity, in particular, brought with it colonization and the impact of Western religion; diseases and slavery reduced the number of indigenous peoples dramatically. In certain areas, indigenous faiths were wiped out. This, combined with major programmes for the conversion of indigenous peoples by both religions (and the slavery practised by both Christianity and Islam) has led to the virtual eclipse of indigenous religions in vast areas of the world. (For areas in which they have survived see pages 34–35, Traditional Beliefs.)

The rise of Marxism/Leninism and its associated secularism had a formidable impact on the practice of Buddhists, Christians, Jews and Muslims throughout the world. Communist states contributed more than any other cause to the persecution of religions and loss of religious power. The collapse of communism in the former USSR and Eastern Europe in 1991 has resulted in a remarkable revitalization. Religion is being rediscovered even where communism is still powerful and asserting official disapproval, as in China and North Korea. Religious revival is not just an attempt to return to the past, for the faiths have all been changed by their experience. The survival of religious teachings, practices and beliefs where no outward sign is permissible (for example, Albania, 1967–91, or China during the Cultural Revolution, 1966–76) has taken many by surprise.

Sources

David B Barrett, George T Kurian, Todd M Johnson, *World Christian Encyclopedia*, vol 1, 2nd edition, Oxford University Press, 2001

Geoffrey Barraclough (ed) *Times Atlas of World History* (4th edition), Times Books, 1995

Roots and Branches

(pages 18–19)

The roots of religious change and division are diverse. They do not necessarily stem from theology and may owe much to political, economic, and social issues. When a religion moves across borders and cultures, it may have to adapt to new conditions. New forms may develop, different from, or even at odds with, more traditional interpretations. Buddhism, for example, spread from its original homeland in India, into China, Tibet and Japan. Many new schools of Buddhism arose that were often radically different from earlier traditions, but made sense within the new cultures.

Political and ethnic differences may also create new schools, divisions or denominations. The success of the Protestant Reformation in 16th-century Europe owes as much to the rise of nationalism, expressed through the evolution of such 'national' churches as the Church of Sweden or the Church of England, as it does to substantial theological disagreement.

Some divisions have arisen from the desire of people to make a religion their own and from resistance to religious authority. Local expressions of an international religion may develop, as with the many African indigenous churches such as the Church of Zion in South Africa, and even be encouraged as a source of vitality.

In countries where there are different religions or divisions, details are given only for those forming the majority, and within them only the key traditions, schools and groups have been listed, although there may be myriad smaller ones.

In Buddhism, there are two major divisions: the Southern school, known as Theravada, and the Mahayana tradition of the Northern and Eastern school. Theravada or 'Teaching of the Elders', which dates from the time of the Buddha 2,500 years ago, is considered to be the oldest school, and its holy books, the Pali Canon, the ones closest to the words of the historical Buddha. The Pali Canon has shaped Buddhism in Sri Lanka, Burma, Thailand and much of South-East Asia. Mahayana or 'Great Vehicle' arose around 2,200 years ago, and its teachings have shaped Buddhism throughout China, Korea, Japan, Tibet and Mongolia.

Within Christianity, there were a series of historical and theological periods of division. The Roman Catholic Church and Orthodox Church, originally the western and eastern wings of the same Church, finally split into two sections in 1054. Differences over interpretation of authority and theology brought out tensions between the newer Eastern Roman Empire, based in Constantinople, and the older Western Roman Empire, based in Rome. The 16th-century Protestant revolution reacted against Catholicism and created new denominations. Some, such as Lutherans and Anglicans, essentially continued the old style of Church, with bishops and other Catholic practices, but made the monarch head of the Church, rather than the Pope. Over time, other Protestant denominations developed that rejected bishops and the Catholic theology of the State churches. Presbyterians, Congregationalists or Quakers all sought to develop models of leadership and theology they believed were those practised by the early Church.

Hinduism is not really a single religion, so it is not appropriate to suggest divisions. Hindus often refer to their religion as *sanatana dharma*, or eternal truth, one of its ancient names. There continue to be hundreds of different ways that this is expressed, the two main forms being Vaishnavite (special devotion to Vishnu) and Shaivite (special devotion to Shiva).

In the mid-7th century, Islam divided over the source of religious authority within the religion, into two major divisions: Sunni and Shi'a. The word Sunni derives from Sunna – the code of behaviour, customs and sayings of the Prophet Muhammad and his early companions, contained in the literature of the Hadith. Sunni means 'Followers of the Smooth Path'. Sunnis, who form the majority of Muslims, hold that the first three caliphs were all Muhammad's true successors. They follow one of four main schools of law: Hanafi, Hanbali, Maliki or Shafi, all founded by Muslim jurists in the 8th and early 9th centuries.

The Shi'a believe that Ali was Muhammad's first true successor and the term Shi'a originally referred to the partisans (shi'a) of Ali. There are a number of major sub-groups, the largest being the Ithna'shaariyya (the Twelvers), so-called because they believe that the twelfth imam of the line of Ali was not murdered and did not die when he disappeared aged four years old. They believe he will reappear as the al-Mahdi – the chosen one – who will herald the end of the world. The Zayids are a further sub-group. They were founded by the 5th imam after Ali and are now found primarily in the Yemen. The Ibadiyyah was the first group to split away from the main body of Islam, while claiming to preserve the true teachings of Islam. Based on a group around Khadijah, the Prophet Muhammad's wife and first convert to Islam, they are very strict in their observance, live mainly in the deserts of Arabia, Iraq and North Africa and are almost entirely Bedouin.

Since the map shows only the divisions of the numerically widespread religions, Jewish groups are not listed by tradition. Orthodox Jews, who are in the majority, assert the supreme authority of the Torah and believe that Jewish laws are not open to revision. Moves away from traditional or 'orthodox observance' have given rise to Reform, Liberal, Conservative, Reconstructionist and other forms of Judaism.

There is no major doctrinal division within Sikhism.

The Daoism best known in the West is often called 'philosophical' Daoism after the great mystic writers, Lao Zi, Zhuang Zi and Lieh Zi of the 5th to 3rd centuries BCE. However most Daoism as actually practised comes from what is called 'religious' Daoism, founded by Zhang Dao Ling in the 2nd century CE. Many different schools emerged over the next 1,000 years, although only a few survive in any significant form today. Much of popular Chinese indigenous religion is Daoist-influenced, with a strong overtone of Chinese popular Buddhism.

Sources

E Breuilly, J O'Brien, M Palmer, *Religions of the World: The Illustrated Guide to Origins, Beliefs Traditions & Festivals* (revised edition), Facts on File, 2005

David B Barrett, George T Kurian, Todd M Johnson, *World Christian Encyclopedia,* vol 1, 2nd edition, Oxford University Press, 2001

International Religious Freedom Reports 2006, US Department of State, www.state.gov/g/drl/rls/irf//2006/

Christianity

(pages 22–23)

Christianity is growing across the world, especially in Sub-Saharan Africa, parts of Asia such as South Korea and China, Central Asia and in the lands of the former USSR. Christianity has explicit missionary teachings within the Bible, and within the tradition of the religion. While other faiths have an element of mission, usually seen as being carried out through personal contact and friendship, organizations within Christianity place a structural and financial emphasis on taking their message to other peoples, as well as to non-practising members of their own religion (see pages 52–53, Christian Missionaries).

While there are many small Christian communities in Muslim majority countries, often pre-dating Islam, their numbers are decreasing through migration (as in Syria, Iraq, Egypt or Lebanon). However, there are large numbers of Filipinos, Koreans and Christians from other countries living and working across Muslim heartlands. While Christianity has flourished in some areas, such as Africa, some ancient cultures have been more resistant. India and China, despite centuries of effort, have proved difficult for Christians to penetrate. Although Christians are numerous in densely populated India, they remain a small minority. In China, until very recently, never more than 1 percent of Chinese would call themselves Christian. However, the situation is changing. Although it is difficult to be certain quite how many are now Christians, even the lowest estimates exceed 3 percent.

One aspect of Christianity not often commented upon is its ability to adapt to, and often adopt, local religious practices which would be frowned upon by its more traditional elements. In Sub-Saharan Africa and South America, in particular, Christianity has often fused with indigenous traditions to create localized expressions of Christianity, rooted in local cultures and tradition. Many Christians have also adopted a markedly different Church lifestyle, rejecting the authority of a parent denomination and many of the established aspects of denominational faith and life. In their place often come new authority, structures, names, beliefs and solutions. This is indicated by the bar chart on Independents.

Nearly half of all Christians are Catholics (see pages 36–37, Catholicism), the majority of whom live in developing countries. This socio-economic profile has shifted the Catholic Church in many areas of the world towards what is called a 'preference for the poor', leading to major campaigns for social justice. By contrast, although the majority of Anglicans are also in the developing countries, and amongst the poorest of the poor, and although the Church provides development aid, the reality of the poverty of most Anglicans has not had as profound a theological effect on Anglicanism.

While Christianity is declining in Western Europe as part of the secularization of Europe, it is expanding almost everywhere else – a fact often ignored in Europe, where Christianity has lost much of its social and even cultural force. In Orthodox countries, the experience of persecution has led many to value the Church once again, though the inroads of pluralism are beginning to affect this (see pages 78–79, The Future).

Sources

E Breuilly, J O'Brien, M Palmer, *Religions of the World: The Illustrated Guide to Origins, Beliefs Traditions & Festivals* (revised edition), Facts on File, 2005

David B Barrett, Todd M Johnson, *World Christian Trends AD30–AD2200*, William Carey Library, Pasadena, California, 2001

Census of India 2001

Government of India National Statistics Office Growth Rates 2005

International Religious Freedom Reports 2006, US Department of State, www.state.gov/g/drl/rls/irf//2006

Personal communications: David B Barrett

Islam

(pages 24–25)

Muslims believe that Islam is the religion of all God's prophets from Adam onwards, and see the formal creation of a distinct religion in the 7th century CE as the final form of the religion, explicitly revealed in the Qur'an. Muslims believe that the Prophet Muhammad was God's final prophet and that the revelation given to him has never been corrupted, as it was written down as soon as it was revealed and has been transmitted in the words of the Qur'an. The Prophet Muhammad began to teach the oneness of God in the city of Makkah but, meeting opposition, moved to Medina, where the first Islamic community was founded. Muslim dates are all calculated from this journey (the Hijrah) in 622, which is therefore Year 1 AH (After Hijrah).

Islam is the second largest religion in the world, with 1.34 billion adherents, and is one of the world's fastest growing religions partly through population growth, but also through migration and conversion. While Islam is growing particularly in the areas where it has considerable populations – the Middle East, Central Asia, North Africa and Indonesia/Malaysia – there are also increasingly large communities in North America and Europe. Although in the minority, these growing, active communities began through links created by

Western colonialism, and have continued to expand through family connections or migrant labour from Muslim countries, and by acquiring converts from the host community.

In many Muslim majority countries, Islam is the state religion, and frequently the legal structure of those countries incorporates partial Shari'ah law alongside secular legal systems, or Shari'ah is fully instated as the main system of law. This is a reflection of the centrality of Islam not only in personal life, but in the public and political arenas as well.

The last decade has seen a shift in the profile of Islam, with a highly political, and for some literal, interpretation of Islam coming to the fore. The 20th century gave rise to a number of movements that asserted Muslim identity and independence in response to Western colonialism, but at the end of the century and the beginning of the new one extreme movements within Islam began campaigns of terror, largely in reaction to what is seen as both the decadence and apostasy of some Muslim leaders and countries and the 'threat' of the Christian West (fuelled by military presence or western influence in Islamic countries). Most infamous of all were the attacks in the USA on September 11, 2001. The subsequent invasions of Afghanistan and Iraq by Western allies have only intensified this extreme aspect of Islam, and while this receives much global attention, it is only one of the many strands that exist within contemporary Islam.

Sources

Chris Horrie and Peter Chippendale, *What is Islam, A Comprehensive Introduction*, Virgin Books, 2003

International Religious Freedom Reports 2006, US Department of State www.state.gov/g/drl/rls/irf/2006/

The Pluralism Project – Committee on the Study of Religion, Harvard University www.pluralism.org

Royal Embassy of Saudi Arabia – London UK, Washington DC, USA

Paul Weller (ed), *Religions in the UK 2001–2003*, Multi-Faith Centre at the University of Derby in Association with the Inter-faith Network for the United Kingdom

David B Barrett, George T Kurian, Todd M Johnson, *World Christian Encyclopedia*, vol 1, 2nd edition, Oxford University Press, 2001

David B Barrett, Todd M Johnson, *World Christian Trends AD30–AD2200*, William Carey Library, Pasadena, California, 2001

Census of India 2001

Government of India National Statistics Office Growth Rates 2005

Census of UK 2001

Census of Indonesia 2000

Census of Canada 2001

United Nations World Population Prospects 2004 http://esa.un.org/unpp

Personal communications: David B Barrett

Hinduism

(pages 26–27)

Hinduism is an all-embracing term to describe a vast array of beliefs, deities and traditions. There are, however, certain key features of religious life that show a common root and understanding. Most Hindus would accept a cyclical view of time, as being without origin and without destination. All is reincarnated and all is subject to change, even the very gods themselves. Also, most Hindus believe that the divine permeates everything that exists; everything is divine and therefore mirrors or reflects the image of the divine. This captures the many ways of expressing belief in *sanatana dharma* – the eternal truth, and foundation of all life and reality. Hindus often refer to their overall culture as the Vedic culture. This derives from the Vedas, the oldest Indian sacred books, from which many of the essential tenets of Indian religious and social life derive.

Hindu numbers outside Asia have grown, particularly in the USA, where the number of Hindus almost doubled during the 1990s, and Canada, which saw an 89-percent increase between 1991 and 2001. Many young professional Hindus migrate to take up career opportunities, but Hindu families often retain strong links with India. Most Hindu communities abroad co-exist peacefully, but occasionally tension surfaces. In Uganda in 1972, 70,000 Asians were expelled by dictator Idi Amin, and many migrated to the UK.

The spread of Hindu ideas has been considerable throughout Europe, North America and Australasia. Hindu gurus have travelled to the West in response to the coming of Christian missionaries to India; and Westerners have been profoundly affected by visiting India and by translations of Hindu classics such as the *Bhagavad Gita* or the *Upanishads*. Hindu belief, imagery and philosophy offer a very different form of religious understanding from that practised in the West. It has created a wave of groups inspired by Hindu philosophy and practice – ranging from overtly religious movements such as the Ramakrishna Vedanta Mission, to yoga and Hindu meditation. This has led to a revival of devotional Hinduism amongst many Hindus as well as attracting converts from other religions and cultures.

The Jains are a separate religion, the founder of which, Mahavira, broke away from the prevailing norms of Vedic culture in the 5th century BCE. The Jains practise *ahimsa*, non-violence, and their religious orders of monks and nuns live in such a way as to minimize harm to any living being. The most devout wear a gauze mask over their mouth to prevent them accidentally swallowing any insect, and they brush the path before them to remove ants and other insects. They believe that Mahavira was the last in a line of 24 *tirthankaras* – ford-builders between this world and the spiritual world.

Sources

David B Barrett, George T Kurian, Todd M Johnson, *World Christian Encyclopedia*, vol 1, 2nd edition, Oxford University Press, 2001

David B Barrett, Todd M Johnson, *World Christian Trends AD30–AD2200*, William Carey Library, Pasadena, California, 2001

The Pluralism Project – Committee on the Study of Religion, Harvard University www.pluralism.org

Paul Weller (ed), *Religions in the UK 2001–2003*, Multi-Faith Centre at the University of Derby in Association with the Inter-faith Network for the United Kingdom

ISKCON, London, UK

United Nations World Population Prospects 2004 http://esa.un.org/unpp

Statistics South Africa, mid-year population estimates 2005

Census of India 2001

Government of India National Statistics Office Growth Rates 2005

Census of UK 2001

Census of Canada 2001

Census of Nepal 2001

Personal communications: Richard Prime

Buddhism

(pages 28–29)

Buddhism originated in northern India, founded by Siddhartha Gautama, known as the Buddha – Enlightened One – who lived in the 5th and 6th centuries BCE. For over 1,000 years it flourished there, sending out teachers south and west to Sri Lanka, Burma, Thailand, Cambodia and Vietnam and north to Tibet, China and Japan. It then declined in India itself under the impact of Hinduism and Islam in the 10th to 12th centuries CE. Teaching that life is suffering and that attachment to suffering causes rebirth, the Buddhist path helps the individual to attain wisdom, break the bonds of suffering, and ultimately cease to be caught up in the cycle of birth and death.

In the 20th century, Buddhism was subjected to more sustained persecution than at any previous time. Mongolia, a majority Buddhist country, became the second communist state in the world in 1924. Communism took over China and then parts of South-East Asia from the end of the Second World War until the 1970s. Thousands of monasteries were destroyed – Tibet lost virtually all its monasteries – and most monks and nuns were forced out. In 1930, there were an estimated 738,200 monks and nuns in China; by 1986, there were only 28,000. However, Buddhism is now seeing a remarkable revival in Mongolia and Cambodia, even in countries such as China, Laos and Vietnam, where religious activities are monitored or restricted. The number of Buddhist monks, nuns and lay people is on the increase.

Buddhism is the religion of the majority in Thailand, Cambodia, Burma and Sri Lanka. It is once again the majority religion in Mongolia, as a result of the political freedoms of the 1990s. There are small pockets of Buddhists in Russia, in the republics of Buryat, Tuva and Kalmykia, who claim their Buddhist heritage from Mongolia. Buddhism has gained strength in Indonesia, South Korea, Nepal, India, and China (though religious freedoms for Tibetan Buddhists are restricted and strictly monitored).

Buddhism is the state religion of Thailand, Cambodia and Bhutan, a Himalayan kingdom that only recently opened itself to the outside world. There is likely to be a small Buddhist temple in most villages in Buddhist majority countries, and in the rainy season, when young men take temporary ordination, the numbers of monks in South-East Asian countries can double. The Burmese government is not anti-Buddhist, although increasing political and economic discontent has led to anti-government demonstrations and there is friction between monastic communities and the government.

In Japan, both Buddhism and Shintoism – the indigenous religion of Japan – are followed by the majority of people. Different temples are attended for different rites. Membership of Buddhist sects runs to 56.2 million. Due to traditional family affiliation, many people claim membership of both religions, whether or not they are practising, which makes it difficult to gauge the actual number of practising Buddhists.

In India, the number of Buddhists has increased due to conversions, led originally by Dr B R Ambedkar, from the Dalit or 'untouchable' castes (known as 'scheduled castes' since 1935). There has also been an influx of Buddhists fleeing from Chinese state oppression in Tibet. The spiritual leader of the Tibetans, the Dalai Lama, lives in exile in Dharamsala, in the north of India. While Buddhism is not under threat in most Indian states, Buddhists in Ladakh are cut off from Buddhists in Tibet, and the Muslim population is increasing. Buddhists in Bangladesh are also in difficulty, due to pressure from settlement by majority Muslim Bengalis.

Over the last 100 years or so, Buddhism has spread throughout Europe, North America and Australia. Originally known only through translations of sacred texts and the commentaries of scholars and philosophers, since the 1960s Buddhist influence has increased as a result of great numbers of young people travelling to Buddhist countries. There is an increasing number of Buddhist centres and monasteries outside Asia, but it is virtually impossible to estimate the number of Westerners who practise Buddhism. Relatively few have joined organized religious orders. Many people find some aspect of Buddhist teaching, practice or belief helpful in their religious development, though they may not describe themselves as Buddhist.

Sources

Department of Religion and Culture, Central Tibetan Administration, Dharamsala, 2005

Shukyou Jiten (Religion Dictionary), Gyosei, Japan, 2005

Religion Data Book, 2002, Department of Religions, Ministry of Education, Thailand

Office of the State Council of the People's Republic of China, 2005 White Paper on Freedom of Religious Beliefs

Religious Affairs, National Front for Reconstruction, Lao PDR 2003

Dharma Centre, Republic of Kalymkia, Russian Federation, 2005

Taiwan Government Information Office.

Royal Government of Cambodia, Ministry of Cults and Religious Affairs

Korean National Statistics Office 2003

Censuses of India , Malaysia and Nepal 2001

Government of India National Statistics Office Growth Rates 2005

World Buddhist Directory, www.buddhanet.net

David B Barrett, George T Kurian, Todd M Johnson, *World Christian Encyclopedia*, vol 1, 2nd edition, Oxford University Press, 2001

The English Sangha Trust, Trust Secretary Neil Hammond

Personal communications: Professor Ian Harris, Dr Patrice Ladwig, Tara Lewis, Dr Anna Lushchekina, Mike Shackleton, Guido Verboom

Judaism

(pages 30–31)

While numerically Jews have remained a small group, the impact of their teachings and, in particular, of the Hebrew Bible on the world has been of great significance. Both Christianity and Islam sprang from Jewish understanding and scripture.

After the Exodus of the Jews from Egypt, led by Moses, and the 40 years of wanderings in the desert, the Israelites made their home in the land of Canaan. At first they were ruled by prophets, known as Judges. Later they were ruled by kings, beginning with Saul, followed by David and his son Solomon. After the death of Solomon, the kingdom of Israel split into Judah and Israel, and both were eventually conquered by foreign powers. In the centuries that followed, the Israelites endured exile in Babylon, and although many returned to their homeland, they experienced further periods of invasion. In 63 BCE the Romans conquered the land and named it Palestine. The Jews rose in revolt against Roman rule in 66 CE, but were defeated. The Temple in Jerusalem, rebuilt after the Israelites returned from exile, was finally destroyed in 70 CE. Jewish communities were already widespread in the Middle East, but after the Roman defeat Jews were scattered more widely throughout the Roman Empire. Despite persecutions and restrictions, in time Jews established significant communities in Spain and North Africa, Russia, Persia (modern Iran) and even India and China. The Jewish Diaspora is the name given to the dispersion of Jewish people throughout the world.

Over the last 200 years the location and size of the world's Jewish population has been radically altered by mass migration and today most Jews (80.7 percent) are to be found in either the USA or Israel. From the late 19th to the late 20th centuries a major migration took place to the USA by Jews, who were often driven there by attacks and pressure on their communities in Europe and Russia, the most devastating being that of the Nazi Holocaust in the 1940s. The creation of the state of Israel in 1948 provided an opportunity for further migration, Israel's Law of Return offering automatic right of entry and citizenship to any Jew from anywhere in the world. In the early 1990s, the collapse of communism in Eastern Europe and the former USSR allowed tens of thousands of Jews to journey to Israel. Immigration (Aliyah) to Israel continues. In 2004 the total number of Jews immigrating (Olim) was 20,893. From the beginning of January 2005 to the end of October 2005 the number of Olim was 16,954. The number of Olim since the founding of Israel in 1948 to November 2006 was 3,006,967.

People of Jewish origin have varying degrees of personal commitment to Judaism or Israel. In fact, estimates of the number of Jews worldwide tend to vary because 'being a Jew' can have a religious, ethnic or cultural meaning. In religious terms, a Jew is someone whose mother was Jewish, and who lives by the Law of Moses and of the Torah – the Five Books of Moses or the first five books of the Bible. While many Jews with a Jewish mother would describe themselves as Jewish, they may well not be adherents of the religion.

The figures on the map reflect the 'core' Jewish population: those who clearly identify themselves as being religious Jews or have been identified by a member of their household as Jewish. This includes converts as well as those who have informally joined Jewish groups. It does not include those of Jewish descent who have no religion or follow another religion, those who were born Jewish but disclaim being Jewish, or non-Jewish members of Jewish households.

While the Jewish population of Israel is increasing through births and immigration, elsewhere in the world the numbers are static and in places dropping, due to migration, marrying out and general assimilation. Overall, in 2005, world Jewry continued to be close to zero population growth, with an increase in Israel (1.3 percent) slightly overcoming decline in the Diaspora.

Sources

David Singer and Laurence Grossman (eds), *American Jewish Yearbook 2005*, American Jewish Committee Publications

Analysis of the data in the *American Jewish Yearbook 2005* by Sergio Della Pergola

The Jewish Agency for Israel Central Bureau of Statistics, www.jafi.org.il

Sikhism

(pages 32–33)

Sikhism arose from the teachings of Guru Nanak (1469–1539), who tried to fuse the best in Islamic and Hindu teachings and practice, combined with an intense personal revelation of the nature of God. As a result of religious persecution, the Sikhs established their own community and gained a degree of independence within Muslim-ruled India. Eventually, Sikhs began to spread beyond India, especially to countries formerly part of the British Empire.

Worldwide, the number and size of Sikh communities is gradually increasing, as is clear not only from the population estimates, but from the number of gurdwaras – the temple that is the focus of Sikh religious and community life. A community can range from a few families to a few hundred families. The presence of a gurdwara indicates a Sikh community large enough to support both it and its associated institutions. Beyond Asia, in particular in the USA and Canada, there are growing Sikh communities. The number of Sikhs in Canada increased by 89 percent between 1991 and 2001, while the US Sikh community continues to grow, particularly in California, which has the highest Sikh population of any US state.

Within India, Sikhs are migrating to the Punjab in increasing numbers. This is in response to levels of disturbance and tension between Sikhs and the Indian government. In turn, these are partly due to a growth of Hindu nationalism and to Sikh aspirations for their own separate state of Khalistan (Land of the Pure). In 1984, to crush the growing Sikh independence movement, Indian government troops stormed the holiest shrine of Sikhism, the Golden Temple of Amritsar. At the beginning of the 21st century the situation is calmer, but the undercurrent of resentment is still an issue within the global Sikh community, which continues to feel under-represented as a minority community in India. Another focus of concern for today's Sikhs, especially for the diaspora Sikh community, is the question of how to maintain and redefine their identity and values in an increasingly Westernized global community.

Sources

Census of Canada 2001

Census of UK 2001

Census of India 2001

Government of India National Statistics Office Growth Rates 2005

Professor Baljit Singh Moonga, www.columbia.edu/-bsm15/baljit.htm

www.sikhnet.com

United Nations World Population Prospects 2004 http://esa.un.org/unpp

Personal communications: Dr Rajwant Singh, World Sikh Council

Traditional Beliefs

(pages 34–35)

Shamanism is the term that has come to describe a religious belief and practice that is found in some form in traditional societies throughout the world. The term 'shaman' originated with the Tungus tribe of Eastern Siberia and has been used to refer to similar beliefs and practices worldwide. Some argue that Shamanism spread from Siberia and Central Asia to the Russian steppes; from there it migrated eastwards with the movement of peoples over the land bridge linking Siberia and Alaska, and also spread westwards into Europe. Shamanism is rooted in the belief that there are two worlds – a spiritual world and the world of material existence. A shaman possesses an ability to communicate with the spiritual world through an altered state of consciousness.

In many parts of the world, traditional beliefs and cultures survive only in inaccessible or difficult terrain – the last refuge of peoples who formerly inhabited more accessible land but who have been pushed back, whether by aggressive settlers, restrictive government policies, persecution for their beliefs or assimilation into the majority culture. In the Amazon basin, the loss of traditional lands is still going on under pressure from logging and the encroachment of ranching, soya farming and urban centres. In China, many of the 55 officially recognized ethnic minority peoples (there are also ethnic groups that are not officially recognized) live in mountainous areas, where they were driven by the spread of the Han Chinese over the course of the last 3,000 years.

Though traditional beliefs form the majority religion in only a few countries, their influence is still widespread. In China, popular religion is observed in most villages and towns. This is a mixture of traditional beliefs focusing on ancestors, and on the intermingling of the spirit world and the physical world, overlaid with a veneer of Buddhism and Daoism.

In Japan, Shintoism, unlike many traditional beliefs, has a written, rather than oral, sacred literature. It is a separate religion from Japanese Buddhism, but the two co-exist and many people follow both religions. Shintoism, in common with most local traditional beliefs, has neither sought nor gained a following outside its own people.

It has been the spread of Christianity and Islam that has had the most dramatic impact on traditional beliefs over the last 500 years, fundamentally affecting the religious practice of indigenous peoples from Siberia to Africa, from the Amazon to the Australian outback. In some places, whole belief systems and formerly numerous indigenous peoples have completely disappeared. In some areas, people have continued to practise tribal traditions as well as Christianity or Islam, which have either been changed by traditional beliefs or learned to accommodate them. The Catholic Spiritists of Brazil, for example, while clearly Catholics, have incorporated specific characteristics of African religions, originally brought to the region during the period of the slave trade (see pages 40–41, New Departures).

In some African states where Christianity or Islam has become the majority religion, traditional beliefs have mingled with them and have produced idiosyncratic expressions of a religion, usually confined to a single area. The growth of 'indigenous churches' highlights the fusion of certain core Christian ideas with key elements of traditional belief. The Church of the Initiates in Gabon, for example, combines ancestor worship with specific Christian teachings and symbolism. There has also recently been an increased interest in traditional beliefs amongst some people in the West, thereby bringing the insights of traditional peoples into a wider circle of religious awareness.

There has also been a revival in traditional beliefs among indigenous peoples of Australia, Canada and the USA – not always welcomed by the government. In Australia, aboriginal peoples have campaigned vigorously for the protection and return of land sacred to them, such as Uluru (one-time Ayers Rock). While the governments of individual Australian states have agreed to some demands, others are a continuing source of conflict. In the USA and Canada, pride in traditional beliefs and culture has led to a revival of interest, not just amongst Native Americans, but in the wider community (see pages 40–41, New Departures).

Traditional beliefs and ways of life are still under threat – in some cases, at greater risk than ever before. Worldwide, there appears little hope that the traditions, beliefs and close links with the land of indigenous communities will remain intact.

Sources

Canadian Forum on Religion and Ecology www.cfore.ca
Survival International www.survival-international.org
David B Barrett, George T Kurian, Todd M Johnson, *World Christian Encyclopedia*, vol 1, 2nd edition, Oxford University Press, 2001
International Religious Freedom Reports 2006, US Department of State

Catholicism

(pages 36–37)

The Holy Catholic Apostolic Roman Church (or the Catholic Church) is the largest single group within a major world religion. Centred on the Vatican, a self-governing city-state in Rome, it claims the allegiance of 1,086 million believers, and is served by more than 4.2 million people engaged in pastoral activity, including 1.5 million full-time priests, nuns, monks, missionaries lay leaders and catechists. It is the only religious group with full diplomatic representation to national governments as well as to international bodies such as the UN. There are 177 countries that have Nunciatures, diplomatic missions representing the Vatican, including Iraq, Iran, Kuwait, Kazakhstan, Tajikistan, Kyrgyzstan, Mongolia and Russia. If the Vatican does not have official ties with nations, an Apostolic Delegate liaises with the Catholic Church in that nation. The Vatican is the oldest uninterrupted diplomatic service in the world.

Pope Benedict XVI, elected Pope in 2005, is head of the Catholic Church, and the 264th successor to St Peter, the first Bishop of Rome. He is the first German Pope in the history of the Church and, like his predecessor, Pope John Paul II, he stresses traditional conservative Catholic theology, but also leads the Catholic Church in its participation in international issues such as HIV/AIDS support and education, ecumenical and interfaith dialogue and conflict resolution. The Vatican City state has an area of 0.44 square kilometres, with a resident population of under 1,000, but with up to 3,000 additional personnel living outside its boundaries. The Vatican is the centre of an enormous worldwide publishing, TV, radio, and bureaucratic network. A range of specialist congregations and councils, comprising Vatican staff and bishops from around the world, preserve the unity of the Church's teachings. The Congregation for the Doctrine of the Faith is responsible for official statements of Roman Catholic belief. Since the Second Vatican Council of 1962–65, when the Church reorganized and liberalized some of its teachings and organization, there has been increased emphasis on the role of the laity in local decision-making.

Local authority is exercised through the bishops of more than 2,000 dioceses across the world. Bishops gather on a national or regional basis to decide local policies, and visit the Vatican every five years to report to the Pope. A few senior bishops or archbishops are nominated to the College of Cardinals. In June 2006, there were 192 Cardinals, 53 of whom worked for the departments of the Roman Curia within the Vatican. The College of Cardinals is responsible for the election of a new Pope on the death of the incumbent, but cardinals can vote only if they are under 80 years of age at the time of the previous Pope's death.

Sources

David B Barrett, George T Kurian, Todd M Johnson, *World Christian Encyclopedia*, vol 1, 2nd edition, Oxford University Press, 2001

David B Barrett, Todd M Johnson, *World Christian Trends AD30–AD2200*, William Carey Library, Pasadena, California, 2001

CARA (Center for Applied Research in the Apostolate) Georgetown University, Washington USA, http://cara.georgetown.edu

NCEA, National Catholic Educational Association, Washington, USA www.ncea.org

National Catholic Register, www.ncregister.com

www.catholic-hierarchy.org

New Religious Movements

(pages 38–39)

The new religious movements illustrated on the map came into being during the 20th century, which probably saw the most dramatic rise in new religious movements in history, fuelled by cheap air travel, the collapse of colonialism, a rise in interest in 'other religions' and the advent of mass media such as television and radio. While many movements may be local – as with the new religions of Indonesia, or the indigenous churches of Nigeria – some have moved across the world and gained a visible presence far from their place of origin. They are in fact new missionary movements, and their spread into other parts of the world is causing some alarm in areas traditionally adhering to Christianity or Islam. Only a small number, selected because of their international reach and their fame or high public profile, is shown on the map. In fact, despite their own claims and the fears of their opponents, they are still in most cases relatively small in terms of actual numbers.

Every major religion has at some time been a new religious movement, splitting away from older traditions: as Buddhism did from Hinduism, Christianity from Judaism, Sikhism from Hinduism and Islam. What marks out today's major religions from myriad other religions over the centuries is their ability to survive, adapt and flourish, by speaking to the human condition.

The increase in communications in the last 100 years has widely disseminated religious ideas, beliefs and teachings throughout the world, many previously unheard of. There are over 150 translations of the Daoist classic *Dao De Jing*, enabling it to be used as a basis for a whole range of religious thought and belief, often different from its traditional use in China. The Gospels of the Christian Bible now appear in 1,400 languages, leading to a host of varied interpretations, often diverging from the understanding of mainstream denominations.

Most new religious movements have their roots in a major world religion. For example, the Unification Church claims to arise from Christianity; the Brahma Kumaris and the World Plan Executive Council (Transcendental Meditation) from Hinduism. Others see themselves as completely new. The School of Economic Science, for example, arose from a teacher who sought to fuse economics with philosophy, which was later claimed as a link to certain forms of ancient Hindu wisdom. The Scientologists are one of the best known of these groups. They believe that the individual is trapped by 'engrams' – negative experiences in childhood or even in the womb, which cripple and hold people back from full development. Using a system called 'Dianetics', founded by the former science-fiction writer L Ron Hubbard, they claim to be able to free the individual from these engrams and thus ensure a more fulfilled person.

New expressions of an existing religion may develop when it is transferred to a totally different culture. Buddhism, for example, was transformed as it travelled through China, producing new forms such as Chan (known as Zen in Japan) that have become part of the range of mainstream Buddhism. The opposite can happen. Ideas from another religion may be taken back home and developed within a different culture. The older religions have often disowned the activities of people who claim to be followers. The Unification Church of South Korea, for example, claims to be Christian, and much of its language and symbolism comes from Christianity. Indeed, its full title is 'The Holy Spirit association for the Unification of World Christianity'. But its leader, the Reverend Moon, believes he is the new messiah come to complete what Christ failed to achieve – the founding of a Perfect Family. Such dramatic divergence from mainstream Christian thinking means that Christianity rejects the Unification Church.

New religious movements often flourish in already pluralist and often volatile societies, although they do not always travel abroad. In Indonesia, over 60 million people follow indigenous new religious movements, which are an officially recognized category of religion.

New religious movements may be accused of corruption or of abusing people through 'brainwashing' or totalitarianism. But most of them are neither sinisterly manipulative nor destructive. More often than not they are targeted by parents or families who feel they have lost control of, or contact with, their children. Many more young people break off contact with their parents for reasons that may have nothing to do with religion.

A few groups that have been classified as new religious movements are now considered to be new religions, or serious branches of major religions. The Baha'is do not consider themselves to be a sub-group of Islam, but a new religion. They arose from Persian Islam (modern Iran) in the mid-19th century, and by the end of the century had spread well beyond the Islamic world, often being embraced by those who wanted a religion similar to, but not connected with, Christianity. The International Society for Krishna Consciousness (sometimes popularly called the Hare Krishnas) is considered by some as a new religious movement but increasingly is seen as a branch of Hinduism, and as such their numbers have been added to those for Hinduism in this atlas.

Most new religious movements are not the cranks and crazies of the Western popular press. They are usually serious, legitimate and highly organized expressions of religious belief, or re-workings of older religious traditions. Many do not see themselves as having diverged from the main tradition, but as being more faithful to the original founder than the dominant expression of that religion. However, there is always the risk of abuse with charismatic-led or highly controlled religions – although this can be as much a feature of the major religions. One of the earliest known cases of anti-brain-washing techniques being used to prevent a family member joining a religious group occurred in the 13th century. The family of St Thomas Aquinas, wanted him to go into the family business, whereas he wanted to become a monk. They kidnapped him, locking him away for months to try and break his intention. They failed, and the Church gained its greatest medieval thinker.

Source

INFORM (Information Network Focus on New Religious Movements), London School of Economics www.inform.ac

New Departures

(pages 40–41)

The movement and development of religions have been shaped by older indigenous traditions that have migrated with people and, more recently, with ideas and trends such as the growth in the late 20th century of the New Age movement. A key characteristic has been the ability of the original beliefs and rituals to blend with the traditions they have encountered either on their journey or at their destination. This is the case with Santeria, Candomble, Voudoun and Spiritist religions that are widespread throughout the Caribbean and South America, but which also have followers in North America and small pockets in Europe. A wide range of traditional, tribal beliefs arrived in the West Indies and Americas with the people captured and transported as slave labour mainly from West Africa, and it is the traditions of the Yoruba, Bantu, Fon, Ashanti, Kongo and others that flourished with a mix and adaptation of the mainly Catholic Christianity they encountered. The beliefs of indigenous peoples of the Caribbean Islands, Central and South America were also syncretized to create a blend of ancestor worship, veneration of saints, the evocation of spirits and animism with African and Christian rituals, symbols and imagery. In Brazil, it is estimated that up to 8.5 million people are adherents of these religions, but millions more may participate in elements from them from time to time. In Haiti, Vodoun was officially recognized by the state in 2003, and in 1996 was declared the official religion of Benin (but not a state religion in the constitution).

The practices, beliefs and spirituality of the Native American peoples are associated with geographically distinct tribes in North America, ranging across the eastern sub-Artic, the eastern woodlands, the central plains and prairies, and the southwest. A new development has been the creation of a pan-Native American movement called the Native American Church, also known as the Peyote Church, which is active in the USA and Canada with a membership of 250,000 drawn from over 70 Native American nations. While some of the shamanistic practices of the Native American peoples, such as sweat lodges and medicine lodges, have spread out to a wider group in the USA and Europe, most modern western Shamanism was influenced by studies in South America. In the mid-20th century, anthropologists working in the South American rainforests developed a practice called 'Core Shamanism', resulting in the establishment of the Foundation for Shamanic Studies in California. Elements of Modern Shamanism are now found within Spiritual Ecology, the New Age Movement and Paganism.

After the arrival of the first British missionaries in New Zealand, Christianity spread rapidly, and indigenous beliefs and practices were syncretized with various Christian traditions to create new religious traditions. Two independent Maori Churches, the Ratana Church and the Ringatu Church, continue to the present time, and involve 1.7 percent of the population. In a 1996 survey, 60 percent of Maoris in New Zealand identified themselves as Christian of various denominations but within this Christian context, Maoris continue to draw on older indigenous traditions. While the New Age movement has appropriated some Maori religious traditions, this has not been on the scale equivalent to the borrowing from Native American traditions.

In addition to the reinvention of Celtic Druidry and Heathenry, there is also a revival of Baltic Paganism following the collapse of communism. There is evidence that pagan religion survived in the more isolated parts of Eastern Europe, for example, Estonia, Latvia and Lithuania, and these pre-Christian traditions are becoming part of the national identity of these newly emerged countries.

Sources

INFORM (Information Network Focus on New Religious Movements), London School of Economics www.inform.ac

David B Barrett, George T Kurian, Todd M Johnson, *World Christian Encyclopedia*, vol 1, 2nd edition, Oxford University Press, 2001

Non-Believers

(pages 42–43)

The issue of religion can be very difficult for some people. Some are uncertain that they can believe anything specific about God or the divine and thus classify themselves as agnostics. Others reject all belief in God or the divine, defining themselves as atheists. For some, damaging experiences of religious groups or individuals in their childhood have put them off religion for life. In a few countries, as for example in Cuba and North Korea, it is a prerequisite for political position to be an atheist. However, it is not socially acceptable everywhere to profess to being an agnostic or atheist. In a handful of states, such as Iran, it is even dangerous.

In many European states, belonging to a religion no longer carries with it the notion of full adherence to a set of teachings, and there exists a form of agnostic Christianity. Many people, and not just in Europe, would wish to retain moral and ethical insights and codes of religion but abandon overt religious practices. They see themselves as personally agnostic but accept the need for society to adhere to a shared moral code. When adults in the UK have been questioned about religious education, over 50 percent regarded it as important for imparting a moral code, but not so important for imparting religious faith or truths.

Those who identify themselves as atheists often do so in order to make a stand, not just against the beliefs of any given religion, but against the whole apparatus of organized religion. They may take a position that religion is not rational, or subject to reasoned study and analysis, and this may be fused with personal experience of the failings of religion. Some atheists regard religious faith as being a primitive state of mind, one that hinders the fullest development of humanity. Atheism can itself become a belief system, replacing God as the centre or purpose of life. Atheists have been instrumental in founding many international humanitarian agencies, such as the United Nations or the International Union for Conservation of Nature.

The activities of official atheist states, former or current, has caused problems for many atheists. The brutal and forced suppression, or even the eradication, of religion has meant that as a movement rather than a personal stance the term 'atheist' is often regarded with suspicion.

Atheism continues to be the official position of the governments of China, North Korea and Cuba. Here, large organizations are to be found, and impressive numbers are cited. It may be politically correct to be atheist in communist countries, but judging by the collapse of such groups and the numbers of atheists once communism falls, the conviction may not always run deep. China is an interesting example: although officially atheist, with religious interference still widespread, the constitution has restored the right to religious activities so long as they do not endanger the State, and religion in China is experiencing revival in many different forms. This could be linked to the current climate of economic change. In Cuba and North Korea, however, systems have remained largely unchanged for many years, and no such revival is evident yet.

In former communist countries, such as the Czech Republic, official atheism may no longer be as powerful, but considerable numbers still claim to be agnostic. The impact of atheism on religious belief has been considerable, especially in states with a Protestant background. It has weakened links with formal religion, legitimated dissent from religious values and traditions, and furthered secularization. Orthodox and Catholic states would appear to have a stronger hold upon people, even after 70 years of communist rule.

The proportion of people who call themselves humanists and who belong to humanist organizations is small. The nature of these groups varies from state to state. In some countries, they have a strong scientific basis and seek to confront what they see as the illogicality of religion. In others they have a quasi-religious foundation – arising from attempts to provide a human-centred religion in the late 19th century. The American Ethical Union is one example. Yet others arise primarily from a concern for human wellbeing, with an emphasis on the need for moral and ethical codes based on human interests and values, and not shaped by reference to a divine will or purpose. The Humanistisch Verbond Belgie (Belgian Humanist League) is an example of this.

There are many who would count themselves as agnostic, or as having no particular religion, and the number is growing. But the numbers of atheists and humanists worldwide are not sizeable, nor do they show any significant increase in recent years. Beyond this, there is no doubt that the legitimation of doubt, and the quest for common human values unrelated to specific religious teachings, has had a considerable impact even on those who profess a faith.

Sources

International Humanist and Ethical Union, www.iheu.org
David B Barrett, George T Kurian, Todd M Johnson, *World Christian Encyclopedia,* vol 1, 2nd edition, Oxford University Press, 2001
Personal communications: David B Barrett

State Attitudes

(pages 46–47)

State neutrality on religion, or religious tolerance, is undoubtedly gaining worldwide support. In the West, the secularization of elites, a general increase in individualism and personal choice, and the decline in the significance of public religious practices mean that religion is more and more seen as a matter of personal choice – of little concern to the State unless specific religious activities challenge its interests. For example, both Norway and the UK have formal state religions, yet over the last 100 years the degree of control exercised by the State has markedly declined and a *de facto* division allowed or even encouraged. In Belgium, a staunchly Catholic country, state recognition is now afforded to many different traditions: Christianity as well as Judaism, Islam and even secularism. The links between a religion and national identity are rooted in the past and built up over centuries. In states without a religious history, or where the state itself has been subject to change, the ties with religion are more tenuous. The map shows clearly that few 'new' nations – especially post-colonial states – have formal ties with any one religion. Where they do, such ties are usually with a pre-colonial religion, as in Pakistan or Burma. For the majority of 'new' African states, the concept of mature statehood assumes that religious practice is a matter of individual choice, so long as it does not endanger others or state interests.

Yet in many countries of the world, the concepts of State and religion are still intimately intertwined. In many Muslim countries, it is inconceivable that the head of state should not be a Muslim, or that the government should not follow Islamic law or attempt to rule according to Islamic principles. This stems from Islam's belief that the Qur'an and Shari'ah contain all that is necessary for the proper governing of a state.

The State may exert its authority over religions and also over its people, through either recognition or oppression. The State may also seek to protect the dogmas and certainties of its ruling group and elite. Much of the conflict between Marxist-Leninist regimes and religious communities comes from the fact that Marxism-Leninism was to all intents and purposes a belief system itself – and one that was particularly threatened by other belief systems.

A few countries, having experienced religious conflict, seek to hold the religions in balance. Indonesia for example, increasingly troubled by religious tensions, has for many years recognized six religious groups – Islam, Buddhism, Hinduism, Catholicism, Protestantism and new religious movements – and recently, after anti-Chinese riots, added Confucianism.

In a number of states during the 20th century, political revolutions have broken once-close ties with a majority religion: between Russia and the Orthodox Church, for example, and Cambodia and Theravada Buddhism. This radical change often instigated a period of persecution, for both the majority religion and other religions too. In both cases, the old dominant religion still retains its majority status. While Buddhism in Cambodia has been accorded state religion status again, this has not been the case with the Orthodox Church in Russia. Pluralism often becomes a feature of state thinking after political revolutions and there is a tendency to move from a single religion to recognition and honouring of all mainstream traditions, in so far as they do not threaten national security or disturb community relations.

Sources
BBC News Country Profiles http://news.bbc.co.uk/2/hi/country_profiles
International Religious Freedom Reports 2006, US Department of State, www.state.gov/g/drl/rls/irf//2006
Country Reports on Human Rights Practices 2005, US Department of State
www.state.gov/g/drl/hr/c1470.htm
David B Barrett, George T Kurian, Todd M Johnson, *World Christian Encyclopedia*, vol 1, 2nd edition, Oxford University Press, 2001

Christian Finance

(pages 48–49)

Before 1900, 90 percent of Christian giving was channelled through churches and denominations and 10 percent supported Bible societies and missionary societies. The 20th and 21st centuries have seen the emergence on a massive scale of over 40,000 distinct new parachurch agencies whose finances are independent of the churches. By 1980, 36 percent of all Christian giving to Christian causes was by-passing the churches and denominations. By 1995 this had risen to 60 percent and continues at this level.

The most regular annual surveys of giving come from the USA, where 81 percent of all giving is by individuals. A survey of giving trends of church members in 29 composite US denominations in 2004 indicates that financial giving to Churches and Christian benevolent causes was 2.56 percent of income (amongst these largely Protestant denominations a high proportion was, in fact, given directly to support the internal work of the congregations). The survey reveals that although the total giving by church members as a percentage of income has fluctuated only slightly in recent years, there has been a steady decline from 3.11 percent in 1968.

Christian giving in North America to religious and secular causes through all denominations does, however, exceed $100 billion annually. Just over 90 percent is for Christian causes and the remainder for secular or non-Christian causes. While this reflects Christian commitment to charitable giving, it also reflects above-average annual income. While Christians with lower incomes in Latin America, Africa and Asia give numerically smaller amounts, the percentage of giving related to income is not necessarily lower and must also take into account giving through time, skills and resources. This ranges from congregations building their own churches, to the running of community church projects. But Christians in the poorer nations of Latin America, Africa and Asia also have some of the most dynamic forms of Christianity, reflected in the growing number of indigenous Churches and independent Church movements. Churches such as the Zion Christian Churches of South Africa receive more than $200 million annually from their members, while the Assemblies of God in Brazil receive more than $1 billion in annual donations from their congregations.

Sources
David B Barrett, Todd M Johnson, *World Christian Trends AD30–AD2200*, William Carey Library, Pasadena, California, 2001
The State of Church Giving through 2004, www.emptytomb.org
Personal communications: David B Barrett

Religious Education

(pages 50–51)

The place and role of religion within a state can sometimes be best understood by looking at the role and status of religious education within state schools. Traditions of secularism, atheism, of formal links between religions and governments, or the increasing move to a pluralist approach towards the role of religions can all be seen mirrored in the status of RE. However, this can also work in reverse. For example, the USA is a devoutly religious country with a very high proportion of the population attending places of worship on a regular basis, yet RE is banned from schools. In Europe, where religious observance has been declining for decades, the Council of Europe wants every state in Europe to have formal RE.

Every education system in the world owes its origins to religious schools, and to the concern for education within the major religions. Religions have always made education a high priority: to ensure continuity of the religion, to pass on stories and beliefs, and to maintain the moral and social wellbeing of the community. Thus, many of the world's greatest universities and centres of learning have religious foundations, such as Oxford and Cambridge in the UK, Harvard in the USA and Al-Azhar University in Cairo, Egypt. Over the last 100 years or so, states have generally set out to wrest control of education away from religion. In 19th-century Germany, for example, Bismarck's nationalist goals for unification were strongly objected to by the Churches. In the UK, the State's policy of acquiring control of education was part of the growing secularization of national institutions.

In some countries, religious education is a state subject, taught by qualified teachers, trained largely by the State or in accordance with its requirements. In some states, as in Sweden, its purpose is to inform about the role of religion, seen as central to a mature understanding of culture and society. In others, it may be part of an agreement with the majority religion, to ensure that children become familiar with the religion of the culture (as in Bahrain, a Muslim majority state, and the Dominican Republic, where the majority of people are Catholic). In increasingly plural societies, for example in South Africa, religious education has been broadened to include teaching not just about the majority religion but other religions as well.

In a number of countries, religious education is not a formal part of the state curriculum, but religious teachers are allowed to hold special lessons on school premises, which pupils may opt to attend or not. In India, for example, religious education is optional at the choice of parents or children, and is offered out of normal school hours, even in schools still under the authority of a church. In other countries, religious education is not permitted within the school curriculum. In the USA, this has been an historic position because of the separation of Church and State in the US constitution. In the early 1990s it is beginning to change in certain states, such as California. In China, religious education is not permitted because the State is officially atheist.

In the new states of the former USSR and Eastern Europe, the role of religious education is under scrutiny, as is education in general. In Poland, debate on the role of the Catholic Church in the country as a whole has centred on the form of religious education to be taught in state schools, while in Russia fierce debate has taken place over the role of RE in a country that currently has no state RE, but where some feel the Orthodox Church should have a role in state education.

In Western Europe a syllabus may include the study of religions not represented within the population at large. However, teaching of Christianity, the majority religion, may still predominate in some countries. In the world in general a broader outlook is developing within religious education, and states are tending to move away from narrow and confessional-based syllabuses. Religious education is more and more judged by educational standards relating to child development, rather than by religious standards of conversion or of nurturing children into a specific religion.

Sources

David Lara, *Libertad religiosa y educación religiosa escolar*, Bogota: PUJ, 2006.

Personal communications: Professor John Hull, Professor Emeritus of Religious Education, University of Birmingham, UK; Professor Robert Jackson Director of the Warwick Religions and Education Research Unit, University of Warwick, UK; Dr Gunnar J Gunnarsson, Dr Pille Valk, University of Tartu, Estonia; Dr Chrissie Steyn, Dr Peter Schreiner, Comenius Institut, Munster, Germany; Dr Jose Luis Meza and Dr Roseanne McDougall

Christian Missionaries

(pages 52–53)

The Christian missionary movement is as strong, if not stronger, than it has ever been. It has 4,100 missionary organizations, over 419,500 full-time career missionaries and an annual expenditure of $15 billion. It is far removed from the 19th-century stereotype of the missionary, dressed in white and paddling up the Limpopo, or the notion of going to a 'heathen' country to bring about conversions.

Christianity has always been a missionary religion – the Apostle Paul was its first major missionary figure – but in the 21st century, just 1 percent of all Christian missionaries work with non-Christians. Almost all go to Christian majority countries in response to requests from local churches, a principal known as 'partnership'. A massive amount of time, energy and finance is spent on 'home' missions – and 70 percent of missionaries work in their home country. Further, the total impact of foreign mission is difficult to calculate, since large numbers of Christians work abroad for non-Christian organizations and engage in part-time missionary work. Others go out under the auspices of groups not officially described as missionary organizations.

As well as preaching and teaching, foreign missionaries today work on agricultural schemes, housing projects, schools, hospitals, environmental programmes, counselling programmes, computer materials, information systems and church organizing. Most missionaries work within a major denomination, and some as itinerant or independent evangelists. Some use TV or radio to spread the gospel, some still travel on foot, by camel or by bicycle.

The largest numbers of missionaries still come from the traditional mission-sending countries, although former 'mission field' countries are sending numbers to work abroad, for example, South Africa sent out 7,000 missionaries, Nigeria sent 2,500 and the Philippines sent 2,000.

Since the collapse of communism there has been a dramatic increase in foreign missionary activity in the new states of the former USSR and in Eastern Europe. In 2005, Russia received 19,000 missionaries. Many work with the established Churches – Orthodox, Catholic, Protestant and Independent – but some missionary groups do not view Orthodox Christians or Catholics as 'true' Christians, and have caused widespread dissent. The Russian Churches are concerned that some missionaries view Russia as a non-Christian country.

Protestant missionaries are numerous in other areas where Orthodox or Catholics have traditionally been in the majority, for example, in Central America and the north of South America. They are 'converting' Catholics to the Protestant expression of Christianity – fundamentally altering the religious, social and political balance.

Sources

David B Barrett, George T Kurian, Todd M Johnson, *World Christian Encyclopedia*, vol 1, 2nd edition, Oxford University Press, 2001
David B Barrett, Todd M Johnson, *World Christian Trends AD30–AD2200*, William Carey Library, Pasadena, California, 2001
Personal communications: David B Barrett

The Word

(pages 54–55)

Christian scriptural texts and passages have always been recognized as having enormous evangelizing power. Jesus himself in his ministry constantly quoted Hebrew scriptures, saying 'It is written...' or 'Have you not read....' The Christian Word has been spread through the written words of scriptures; the major way in which scriptures of all religious traditions have been communicated down the centuries is orally.

The Christian scriptures are disseminated orally in many ways, including Bible readings, Bible memorization, Bible advertising, Bible quotation and Bible scholarship. To this is added the whole ministry of writing out scripture verses by hand – usually for person-to-person correspondence, authoring manuscripts and journalism. It also includes printing copies of the Bible and arranging their mass distribution through commercial sales, subsidised or free distribution.

The Bible was first printed by Johann Gutenberg between 1454 and 1455 in Mainz, Germany. This Gutenberg Bible contains the Latin version of the Hebrew Old Testament and the Greek New Testament, and was the first substantial book printed with moveable type. It is thought that up to 180 copies were made and sold mainly to churches and monasteries. In the 16th century the New Testament was printed in German and English, and in 1535 the first complete Bible, known as Myles Coverdale's Bible, was printed in English.

The translation of the full Christian Bible is a complex process, which is why complete translations are available in only 426 languages, as compared with the 2,403 languages into which portions or single books of the Bible have been translated. The majority of translations of the Bible are made under the auspices of the United Bible Societies and agencies, who distributed 372.6 million scriptures in 2005, including 24.3 million Bibles. In most countries these were either sold at subsidized prices or given free. There are currently more than 600 scripture translations underway in 495 unique languages that are receiving technical help from the translation consultants at United Bible Societies, who also give financial support to many of these projects. In addition, there are scores of other organizations worldwide, from commercial publishers to The Gideons International, which produces and places free copies in hotels, conference centres, prisons, educational institutions, hospitals, amongst military personnel and in other public places. This feature of free placement of scriptures is also being taken up by other religions, such as Islam and Buddhism.

Sources

David B Barrett, George T Kurian, Todd M Johnson, *World Christian Encyclopedia*, vol 1, 2nd edition, Oxford University Press, 2001
David B Barrett, Todd M Johnson, *World Christian Trends AD30–AD2200*, William Carey Library, Pasadena, California, 2001
United Bible Societies, www.biblesociety.org
The Gideons International, www.gideons.org
Personal Communications: David B Barrett

Christian Broadcasting

(pages 56–57)

Religious broadcasting is rapidly growing within many world religions. Broadcasting transcends national boundaries and has special appeal for certain religions as a means of sending their message worldwide. For those who listen to broadcasts it can also re-affirm faith or open up a faith tradition afresh, and for some it is the only contact they may have with a chosen religion, particularly in countries where there are political or cultural restrictions on religious practice.

Christians control more airwaves than any other religion. This reflects partly the traditional emphasis on 'the word', and partly the wealth of many Christian majority countries. It is enhanced by the universal role of the English language. Christian broadcasts reach every single country in the world and have close to 2 billion listeners, 30 million of whom are not members of any local church. They are transmitted by 4,000 Christian-owned and -operated stations, more than half of which are in the USA or Brazil, and by state-operated facilities such as the BBC World Service. Many secular commercial or government media have a religious broadcasting component. In the UK, the BBC produces regular religious programmes as part of its commitment to broadcasting at no charge, as do independent TV channels.

There are hundreds of Protestant international radio stations, such as AWR Radio of Costa Rica (Adventist World Radio), which broadcasts from Chile to Puerto Rico and is linked to a further 50 other radio stations via its network in Brazil. In the USA, the North American Mission Board of the Southern Baptist Convention runs and operates nationwide programmes through FamilyNet television and radio. The major Roman Catholic station, Radio Vatican, is powerful enough to reach the whole world, and sends out broadcasts in 40 different languages by 200 journalists in 61 countries. Broadcasters are continuing to increase their audience figures via internet users, making it possible to listen online or download and store broadcast material produced by full-time organized agencies and by part-time lay people. The proliferation of personal computers has enabled internet broadcasting and its associated DVDs, books, music and readings to expand on a huge scale.

Some governments try to block Christian broadcasts, but cannot do so effectively. However, 31 states at least nominally ban all internal broadcasting of the Christian message.

Sources

David B Barrett, George T Kurian, Todd M Johnson, *World Christian Encyclopedia*, vol 1, 2nd edition, Oxford University Press, 2001

David B Barrett, Todd M Johnson, *World Christian Trends AD30–AD2200*, William Carey Library, Pasadena, California, 2001

Vatican Radio www.vaticanradio.org

Adventist World Radio www.awr.org

Personal communications: David B Barrett

Aid and Development

(pages 58–59)

While never a formal part of Christian teaching, the notion of giving some regular amount of personal income to the Church, if only for a favoured charity, is to be found in all branches of Christianity. There has also arisen a tradition of giving aid for overseas development, partly as a result of the 19th-century missionary movement, partly because of continuing ties between the churches of the post-colonial world.

Most mainstream Christian Churches in the West are now able to use formal ecumenical church channels for overseas aid – one of the real successes of the Ecumenical Movement. Protestant churches work through the development arms of the World Council of Churches and agencies set up by national Churches – such as Christian Aid in the UK. Catholic churches are highly organized, with central fundraising programmes in 128 countries linked through the Vatican-based Caritas International.

In some countries, funds are allocated not just from personal donations from individual Christians or local churches, but from a church tax levied by the government. In Germany and Sweden in particular, Church aid agencies benefit from these taxes. Other charities rely mainly, or entirely, on direct donations. The way aid is given differs, and the agencies represented on the map are typical of the broad range of mainstream aid agencies. They usually choose projects, and channel funding through local churches abroad.

The agencies do not fund Christian groups alone. In fact, most have funding criteria that forbid consideration on the basis of religious affiliation. This being said, some aid agencies, big and small, especially from conservative Protestant churches, were created specifically to fund primarily Christian groups. Such groups tend to have a stronger link between aid and evangelization. Christian aid agencies tend to favour projects that enable recipients to achieve self-reliance – through education, practical projects and the establishment of small-scale industries and welfare. While the largest Christian agencies also respond to major emergencies, such as floods or famines, their efforts are mainly directed to long-term solutions of global poverty, rather than on short-term emergency aid, as can be seen in the pie-charts.

The Five Pillars of Islam require Muslims to give or pay 'zakat', an obligatory tax on income, once a year. Zakat is levelled at a 2.5 percent minimum of personal capital wealth, after all basic necessities have been met. 'Basic necessities' are usually understood as being able to feed, clothe, house and educate those within the family. There is no legal force or administration to oversee the giving of zakat; it is left to personal honesty and it is often supplemented by further personal acts of charity. Both are given through the local mosque, and distributed locally or further afield through informal networks between mosques and Islamic organizations. A vast amount of Islamic aid is therefore untraceable or unrecorded in any accessible way.

During the 1980s, in part in response to similar

Christian programmes, Muslim groups began to organize formal Islamic aid agencies, especially in the West, where the informal networks are less strong. Aid is primarily given to Muslim states or to Muslim groups in non-Muslim states. Funds are given to develop local skills and projects, such as cooperative businesses in areas of high unemployment, or farming in areas of devastation. They are also given for emergencies, such as emergency relief to Indonesia after the 2004 tsunami, or to communities in areas of conflict such as Iraq and Afghanistan. Only certain large Muslim agencies are shown on the map. Most Muslims still prefer to give through their own mosques, which means that the total Muslim aid channelled through Islamic agencies is, so far, not very large.

One side effect of the 'War on Terror' by the USA and others has been a major curtailing of the ability of USA-based Muslim charities to function. Some have had to close, others have had to spend increasingly large sums of their income on management in order to fulfil criteria for sending funds overseas. The accusation of supporting terror has also curtailed donations. This situation has meant that many projects that were funded by Muslim agencies and charities, particularly in Africa, have had to close, increasing the poverty of the poorest of the poor. Another form of aid, however, is given directly by Islamic states, in particular, Saudi Arabia. In this way, large sums are invested in development projects, usually in the poorer Islamic states or Islamic communities in countries such as Tanzania or Kenya. Such funds tend to be administered by Muslim religious missionaries and are usually for education, health and welfare programmes.

Sources

Christian Giving

Catholic Agency for Overseas Development (CAFOD) Financial Statements 2004/5 www.cafod.org.uk/

Catholic Relief Services USA 2005 Annual Report, www.crs.org

Christian Aid, Annual Report 2005/06, www.christian-aid.org.uk

DanChurchAid www.dca.dk/sider_paa_hjemmesiden/home

Evangelischer Entwicklungsdienst (EED) www.eed.de/fix/files/doc/eed_Bericht_05_06_eng.2.pdf

International Orthodox Christian Charities www.iocc.org/pdffiles/taxforms/2005_annualreport.pdf

Lutheran World Relief www.lwr.org/about/docs/2005ar.pdf

World Vision International www.wvi.org/wvi/pdf/WVI2005AnnualReport.pdf

Muslim Giving

Aga Khan Development Network www.akdn.org

Human Appeal International www.humanappeal.org.uk/about/annualreport.htm

Islamic Relief Annual Report and Financial Statement, year ended 31 December 2005 www.islamic-relief.com

Life for Relief and Development http://www.lifeusa.org/site/DocServer/financial_report_2004.pdf

Mercy-USA http://mercyusa.org/2005AuditReport.cfm

Muslim Aid Annual Review 2005 http://ramadan.muslimaid.org

Islamic Law

(pages 60–61)

Islamic law is founded upon the Qur'an, seen as the infallible word of God, and upon the Hadith, the sayings, actions and silent approval of the Prophet Muhammad. God is seen as the supreme lawgiver, and his laws are deemed to be for the whole of creation, not just for human beings. Throughout Muslim history, issues of interpretation have arisen, necessitating debate and discussion. In arriving at a legal decision, there has to be in theory a consensus of opinion by every scholar, but in practice it tends to be by majority consensus. This is called Ijma – literally 'agreeing upon'. The Qur'an, the Hadith and Ijma legitimize Shari'ah.

It is through Fiqh, Islamic jurisprudence, that Islam covers all aspects of human behaviour, which fall into five categories: *haram* – forbidden, *makruh* – discouraged, *mubah* – neutral, *mustahabb* – recommended, and *fard* – obligatory.

There are four major schools of law within the Sunni tradition, and Sunni Muslims all belong to one of these: Maliki, Shafi, Hanafi or Hanbali. They are named after the 8th- and 9th-century jurists who endeavoured to clarify and set down the law. According to Sunni law, the ruler of an Islamic state is free to choose the school of his choice, just as individuals have the right to be tried according to the school of their choice. The Hanbali school is the most strict and is not found outside Saudi Arabia, where it is the official state law.

Within Shi'a Islam, which does not follow any of these Sunni schools, the 'ulama' – or council of 12 Shi'a elders – is the central political and legal institution. The judges appointed by the ulama eventually became known as ayatollahs, and formed their own courts. In Iran, the ulama is the supreme governing body and the court of appeal for the faithful. The state enforces the Shari'ah law.

While Shari'ah is binding upon all Muslims, it is officially only binding upon non-Muslims when they live in Muslim countries, and even then, only according to a special formulation. In states that follow Shari'ah, for example, while alcohol is forbidden to Muslims under the Shari'ah, it is permissible for non-Muslims to drink alcohol in private, and in ways that will not cause offence to Muslims.

The application of Shari'ah by Muslim majority

countries differs considerably. Certain states, such as Turkey, Albania, Kazakhstan, Senegal and Niger (which also has customary law courts for family matters) operate a purely secular system. Others, often as a result of colonial influence, have a hybrid system that combines secular, Western legal interpretations and structures with elements of Shari'ah: for example, Gambia combines English common law with customary law and Shari'ah law for matters of personal status and inheritance. Indonesia, with the world's largest Muslim population, has a system based on Roman-Dutch law, and Shari'ah is only applied in the religious courts that deal with personal and family matters. The exception to this is the Indonesian province of Aceh, where full Shari'ah courts were formally established in 2003. Yet other countries have Shari'ah as the total legal code for the country, and strictly enforce it. Iran employs the entire Shi'ite version of Shari'ah. In Pakistan, although secular courts exist, they are supervised by Shari'ah courts that have the power to overturn any 'non-Islamic' laws and judgements.

In countries where authority is invested in the monarch, such as Oman, Kuwait and Saudi Arabia, issues not covered by Shari'ah law are decided by the monarch, who may also take into account customary law. Matters related to commercial law, such as oil and gas prices and revenues, are resolved by Royal decrees and codes. In a number of Muslim majority countries the Shari'ah often acts as a focus for the call for Islamic renewal, for example, in Egypt, where the Muslim Brotherhood wishes to enforce Shari'ah as a means of reforming and strengthening Islamic identity.

Sources

Emory University School of Law, Atlanta
Jurist Legal News and Research, University of Pittsburgh
International Religious Freedom Reports 2006, US Department of State www.state.gov/g/drl/rls/irf/2006/
Country Reports on Human Rights Practices 2005, US Department of State
The Government of Afghanistan www.afghangovernment.com
The Law Library of Congress www.loc.gov/law/public/law.html
Law Library Resource Exchange www.llrx.com/library
Chris Horrie and Peter Chippendale, *What is Islam, A Comprehensive Introduction*, Virgin Books, 2003

Faultlines

(pages 64–65)

A frequent complaint against religion is that most of the world's wars have sprung from religious hatred. While the Emerging from Persecution map (pp 66–67) shows this has not been the case in the last 100 years or so, it certainly has often been the case in the past.

Today, the impact and implications of ancient wars, tensions and schisms still profoundly affects contemporary life, politics and society. The historic but largely bloodless split between the Orthodox Church and the Catholic Church, which is formally dated from 1054, still divides Europe into Western and Eastern blocs. The horrific Wars of Religion of the early 17th century, which pitched Catholics against Protestants/Lutherans, and also affected the Orthodox areas of Northern Europe, still mark a division that runs the length of Europe from the Baltic to the Mediterranean.

Faultlines exist where deep divisions have become embedded, and where the suspicions lie just beneath the surface, ready to arise when circumstances become difficult. For example, in Indonesia, the various religious communities have lived quite successfully side by side for much of the 20th century. However, in the last decade of the 20th century, with major social changes taking place as a result of the fall of President Suharto in 1998, religious communities began to fight with each other, often provoked by Muslim extremists attacking churches or temples and then Christian or Hindu groups responding.

A faultline which has developed a new dimension in recent years is that between the French secular state and religion, which, since the Revolution of 1789, has traditionally been between the State and the Catholic Church. With a growing population of Muslims in France, the traditional divide has become a major bone of contention as the French state holds on to its absolute secular status, while the Muslim communities, often very poor and marginalized, argue for religious recognition and rights, such as wearing of headscarves in schools. An old flashpoint is given new significance, but for many of the same reasons of the past.

Sources

Geoffrey Barraclough (ed) *Times Atlas of World History* (4th edition), Times Books, 1995
Martin Palmer (ed) *The Times Mapping History: World Religions*, Times Books, 2004

Emerging from Persecution

(pages 66–67)

The 20th century was the century of ideologies as never before in history: communism, socialism, fascism, nationalism – especially in colonized areas of Africa and Asia – and now, it could be argued, the ideology of consumerist capitalism. Ideologies have always found religions deeply threatening, because the religions espouse different values from the ideologies, and because the religions were in some cases the mainstay of old and often oppressive regimes.

The close links between religion and state, as in Russia pre-1917, Mongolia pre-1924, or Ethiopia pre-1974, meant that when a major and traumatic regime-change took place, the religions were literally in the firing line. The attempt to break up the lands, power and role of religions in so many countries around the world in the last 100 years has led to the most extensive campaigns of destruction against almost all world religions that history has ever witnessed. For example, in Mongolia in 1921 it is estimated that there were 110,000 monks. In the 1930s tens of thousands were murdered, monastic communities were forced to close, places of worship, great monasteries and religious artefacts of all kinds were destroyed. By the time communism fell in 1991, the number of monks was down to a few score.

The rise of nationalism fuelled much persecution. For example, the Armenian massacres in what is now Turkey but was then the Ottoman Empire, in 1915–16, were fuelled by the rise of Turkish nationalism, edged with traditional Islamic suspicion of Christianity. In Mexico, the socialist persecution of the Catholic Church, launched in 1924 and memorably recorded in novel form by Graham Greene in *The Power and the Glory*, sprang from the overweening power of the Church through the centuries, its role in colonization and slavery, and the semi-Marxist theories of the Mexican revolutionaries.

Under communism, religion met an ideological foe head on, which destroyed over a 100 million lives, and wiped out perhaps as many as 90 percent of religious buildings and artefacts in countries such as China or Albania. In Nazism, Judaism faced its gravest ever threat. Nazi fascism took the extreme step of deciding to eradicate all Jews within its reach. Six million died and those that survived only did so because the Allies eventually defeated Germany and her allies. The persecution was founded upon the racial ideology of the Nazis, but was also fed by a centuries-old tradition, linked to Christianity, of anti-Semitism within European culture.

The title 'Emerging from Persecution' has been chosen specifically to indicate that this time of mass persecutions is now over, and the religions are, in almost all cases, emerging. Confucianism is unlikely ever to return to being a significant religion in China or North Korea, but with perhaps that one omission, all the other religions have emerged and are growing stronger. However, none of them is ever likely to return to the position of power they held before being toppled and nearly wiped out by ideologies. Pluralism and a lessened role for formal religions in almost all states affected by such persecution, is now the norm. They may be emerging, but the religions are encountering a very different world from the one they inhabited before this wave overwhelmed so many of them.

Sources

Geoffrey Barraclough (ed) *Times Atlas of World History* (4th edition), Times Books, 1995

David B Barrett and Todd M Johnson, *World Christian Trends AD 30–AD 2200* Pasadena, William Carey Library, 2001.

Shared World

(pages 70–71)

To many it seems only sensible that people of different religions, or of different traditions within one religion, should be able to work together. But the track record of wars and tension in the name of religion bears witness to the intensity of misunderstanding and hatred (see pages 64–65, Faultlines and 66–67, Emerging from Persecution). It was with this in mind that in 1893 the first significant interfaith meeting of the present era took place. A World's Parliament of Religions was held at the Chicago International Exhibition of 1893, to which representatives of all the major religions came. The excitement generated by this meeting was enormous. It seemed like an idea whose time had come.

In the 1960s and 1970s, as interest and contact between religious groups increased, due to expanding opportunities for travel, and the growth of sizeable religious minority communities in Europe, North and South America and Australasia, so interest in interfaith contact and dialogue also increased. While not always able to tackle the more turbulent religious issues, the interfaith movement has nevertheless enabled more amicable relations. In many countries it has led to more sympathetic teaching about other religions in schools.

In some countries, a new interfaith movement has been able to help deal with social, economic or political crises. In the USA, for example, the interfaith movement only really became significant when it emerged out of the crisis of the civil rights movement, and particularly in the aftermath of the assassination of Martin Luther King in 1968. At that time, Jewish, Christian and Muslim leaders sat down to talk and save their neighbourhoods. In Liberia, the interfaith movement was formed in 1992 to enable Christian and Muslim groups to work together in the midst of the terrible civil war that went on for more than a decade. In 1993, Faith Asylum Refuge (FAR) – an interfaith agency dealing with issues of refugees and migration – was set up in the UK alongside other, older, interfaith groups orientated towards social issues. And in former Yugoslavia, during 1992 and 1993, Muslim, Christian (both Orthodox and Catholic) and Jewish groups collaborated in an interfaith network to bring relief to those suffering in the civil war, regardless of creed. That work has continued through the agency of the World Conference on Religions and Peace, which has undertaken practical, social and rebuilding projects, bringing Muslims and Christians of different traditions together.

Increasingly, the need was felt for an international body that would coordinate the interfaith movements worldwide. This has somewhat ironically led to many international interfaith bodies being created. At one point there were at least ten. Four of the most significant, in terms of geographical spread, have been represented here, along with the Council for a Parliament of the World's Religions. However, the heart of the interfaith movement and main activity lies still, as it probably always will do, with the local and national groups emerging from their own social realities, and seeking ways to be faithful together within that context.

The roots of the present worldwide ecumenical movement lie in the missionary experiences of the 19th century. The first major ecumenical, missionary meeting was held in 1910 in Edinburgh, Scotland. In the great vision of that meeting, all Christian Churches would be united in one fellowship. This vision has led to some Churches reuniting, but far more significant has been the discovery of common goals and shared insights and concerns. The old Councils of Churches in many countries have now been replaced by the more dynamic image and title of Churches Together. Nowhere is this more clearly seen than in the struggle against HIV/AIDS. From a hesitant start when some, but only a few, Christian leaders saw AIDS as a divine punishment for homosexuality, the Churches (along with many other religions) have become major partners in the struggle. Their involvement has, however, often been dismissed or trivialized by secular agencies, especially those who find the Catholic stance on condoms problematic, and simplistic barriers have arisen on both sides as a result. Worldwide, the impact of the Churches and of the religions in general on HIV/AIDS far exceeds any other agency or body, yet funding and support from the wider HIV/AIDS industry has often been parsimonious.

The effect of the Churches working so closely together with each other, with other religions and with secular agencies, has been to create bonds of common goodwill and links of common purpose that have begun to overcome centuries of suspicion and resentment.

Sources

United Religions Initiative (URI) www.uri.org (special thanks to Barbara Hartford of URI)
World Conference of Religion and Peace (WCRP) www.wcrp.org
International Association for Religious Freedom (IARF), www.iarf.net (special thanks to Robert Papin)
Monastic Interreligious Dialogue, www.monasticdialog.com plus correspondence and data from Father William Skudlarek of MID
Council for a Parliament of the World's Religions (CPWR), www.cpwr.org

HIV/AIDS

Tearfund: *Faith Untapped*, 2006, www.tearfund.org
Cardinal Javier Lozano Barragan, speech to the UN Special Session on AIDS June 2006.
Data from Rev. Robert J. Vitillo, Special Advisor on HIV and AIDS, Caritas International.
World Conference of Religions for Peace (Study of the response by faith-based organizations to orphans and vulnerable children), UNICEF, 2004, www.wcrp.org/files/RPT-ovc.pdf
Council for a Parliament of the World's Religions, www.cpwr.org
Interfaith movements: Organizations' official websites

Equal Rites

(pages 72–73)

According to most criteria, all the major world religions are essentially patriarchal in their power structures, deities and rituals. Within some religions there are still significant manifestations of the divine feminine, but they are usually subservient – at least in official teaching – to male figures and authorities.

It may be difficult to question why it is almost exclusively men, such as the Buddha, Krishna, Jesus and Muhammad, who are the key figures in their respective religions, but the women's movement has gone some way to focus attention on the male-dominated structures of the major religions. In the case of Christianity, this has focused on the role of women at a sacramental and pastoral level – as priests and ministers. In some Buddhist traditions this has focused upon the right of women to be members of the Sangha alongside monks. In Judaism, it is not only the role of the rabbi that has come under scrutiny, but the participation of women generally in public religious life, while in Islam it has been the role of women in leading prayers, even just prayers for the women's section of mosques.

In theory, at least, the majority of mainstream Protestant denominations within Christianity offer complete equality to women at almost all levels of the Church. All the major Protestant denominations now ordain women – although this is under debate or not yet begun in a few areas or denominations. In the Episcopal Church of the USA, the Anglican Church in New Zealand, the Methodist Church of the USA and the Lutheran churches, there are now women bishops. However, resistance is still strong, and the Churches may be more resistant to change than society at large.

The debate has been particularly fierce within the Church of England and its worldwide body, the Anglican Communion (including, amongst others, Episcopalians in the USA and the Anglican Church of New Zealand). While the threats of a major split have failed to materialize in all instances, the decision to ordain women has created very heated debate, and may lead to small numbers leaving the Church to form traditionalist churches. The pie chart illustrates the very real differences in levels at which women are accepted. But even this doesn't tell the full story. Within many traditions of Christianity that have theoretically accepted women into the priesthood, old recidivist attitudes continue, and women find prejudice expressed in subtle but undermining ways. In particular, the access to top positions is often still blocked – giving what is called the 'stained glass ceiling' effect.

The Catholic Church has as yet to make any significant move on the position of women in the Church, and the late Pope John Paul II officially ruled out any further discussion. However, groups are now being formed in a number of countries, including Austria, USA, UK and Germany, to press for the ordination of women. Alongside this, the actual role of women religious – nuns – has expanded dramatically in recent years due to a shortage of priests. Many nuns, or women who have made lay vows to serve the Church, are now *de facto* running parishes, and to all intents and purposes functioning as parish priests, although without the right to administer the sacraments. The pressure caused by shortages of priests is likely to be one of the major reasons why, in the future, the Catholic Church will allow married priests and then women priests. Many theologians are in favour and have openly voiced their support. During communist persecution of the Catholic Church in former Czechoslovakia, a woman was ordained at least to deacon level, if not to full priesthood. She has now decided to remain silent, but the precedent stands. In similar circumstances, the first Anglican woman priest was ordained in China in 1944. In Russian Orthodox churches, a ceremony already exists for ordaining women as deacons – the final rung before ordination to the priesthood. Yet this has lain unused for centuries, and Russian Orthodox Church opinion is that it is an issue for the future. However, the Orthodox Church of Greece has reinstituted this level of role for women.

While the overall history of Christianity is patriarchal, with some claiming that the early Church had women in senior posts but that they were suppressed, some Churches have been ahead of society in their attitude to women's equality. The Quakers were part of that upsurge of radical thinking in the wake of the English Civil War of the 1640s. The Salvation Army gave full equality from its foundation in 1869, long before women were granted the right to vote in most countries.

In Buddhism, often under pressure from Western Buddhists, the issue of equal rights of participation at all levels by women has led to a revival of interest in women's orders. In some Buddhist countries (notably Tibet and Thailand) the order of nuns died out long ago. The question now revolves around whether any women's orders from places such as China – which has always maintained one – can restore the tradition to other countries. It is a question of continuing authority, but it has also challenged the strong patriarchal assumptions of many traditional Buddhist countries. In Thailand it is seen as being part of the whole debate about the role of women in Thai society and about the extent to which Western values can be merged with traditional Buddhist values.

In Judaism, only the Reform and Liberal wings ordain women as rabbis. Orthodox Jews are opposed. In common with many Christian women, Jewish women scholars are now engaged in extensive re-examination of the Biblical records to show how patriarchy suppressed a more equal relationship between men and women. One aspect of this research has led to a significant debate in both Judaism and Christianity. This is the issue of whether God can be addressed as both Father (He) and Mother (She). This questioning of the assumptions of language about the nature of God has sent considerable shock waves through more traditional sections of the religions, but has also opened up a major debate and exploration of basic notions of the Godhead.

Islam does not have the same sort of debate. For many in Islam, the women's movement as a whole is just another sign of the faltering society of the West. Islam's response has often been to claim that Islam is the true liberator of women. Muslims have questioned what liberation entails, and refer to the harassment that women suffer as a result of ambiguous sexual roles. Many Muslims point to the increase in sexual violence and to the stresses and demands made on women, and ask what sort of liberation this is. Islam sees these as signs of a disruption of a natural order ordained by God. Yet, Islamic law has upheld the right of women to own land and property since the time of Muhammad, whereas women in many countries in the West only fully achieved such rights in the last 100 years. However, in traditional Muslim countries, as in the West, women have been asking for greater freedom and opportunity for equal participation in society. In countries where militant Islam has recently gained strength, changes in women's clothing and behaviour away from the traditional Islamic model have been condemned and prohibited.

Sources

Status of women

Martine Bachelor, 'A women's guide to Buddhism', 2001
The Porvoo Churches Common Statement, the Porvoo Churches, http://www.porvoochurches.org

Ordination of women

Rochester, *Women Bishops in the Church of England?* A Report of the House of Bishops' Working Party on Women in the Episcopate, Church House Publishing, 2004
Ian Jones (ed), *Women and Priesthood in the Church of England Ten Years On*, Church House Publishing, 2004
Christina Rees, *Voices of this Calling*, SCM, Canterbury Press 2002

Environmental Protection

(pages 74–75)

In 1967, an American environmentalist, Lynn White, published a watershed article claiming that the roots of the contemporary environmental crisis lay in Judeo-Christian culture. The command in the Book of Genesis, the first book of the Bible, that humanity should have dominion over all other species, he saw as the root of our use and abuse of nature. White raised a storm of debate about the relationship between religion and the environment, which continues to this day.

In 1986, the World Wide Fund for Nature International invited five major religions to Assisi, Italy, birthplace of St Francis, the Catholic saint of ecology, to discuss and plan for greater religious involvement in the environmental movement. In 1995, WWF helped launch a new organization, the Alliance of Religions and Conservation, which works with 11 of the world's major religions (Baha'is, Buddhists, Christians, Daoists, Hindus, Jains, Jews, Muslims, Shinto, Sikhs and Zoroastrians) to develop environmental programmes and projects through their landholdings, shares, education networks, media outreaches, and moral and spiritual insights. As a result, there are now thousands of religiously based environmental projects.

The religions are significant players because they are important stakeholders in the planet. Between them, they own over 7 percent of the habitable land of the planet; they contribute to, have founded or help in some way 54 percent of all schools worldwide; they have more weekly magazines and journals than the whole of the expanded EU; and their ownership of stocks and shares makes them a major economic force worldwide (see pages 76–77, Ethical Investment and 84–84, Holy Natural).

Although the problems of the environment were first highlighted by secular environmental groups, religious involvement in the environmental issues is undertaken from religious principles and beliefs, and not just in response to secular agendas. They may at first have been seen as 'soft' issues, but the religions are realizing that the wellbeing of the environment is closely connected to fundamental economic and social issues. The environmental crisis has caused a major rethinking within all religions. The religions are showing that they can work side by side on the environment in a way that is unique in history – even in areas of traditional hostility such as Lebanon and the Philippines.

In a number of wealthy countries, the religious aid agencies are now adding an environmental focus to their projects, and emphasizing the link between the environment and development. This has become especially so with the issue of global warming and climate change, which is seen to threaten the most vulnerable parts of the world, the poorest communities and the most fragile environments.

Sources

Alliance of Religions and Conservation, www.arcworld.org
WWF International www.panda.org
WWF UK www.wwf.org.uk

Ethical Investment

(pages 76–77)

Worldwide, the rise of the ethical investment movement has been one of the most astonishing social changes of the late 1990s and early 21st century. The quest for investment in companies and industries that are ethical in their employment, environmentally sustainable, and profitable, has become a major factor in contemporary economics. Ethical banks, investment groups, companies and products are now one of the fastest-growing sectors of the economic world. Ethical foods and products are to be found in every major supermarket in the West, and increasingly in many other areas of the world. The role of the religions in this trend has been an interesting one. Many of the fair-trade groups have their origins in Christian or Muslim religious groups, and the quest for non-interest banking has its foundations in Islam. It has been the growth of ethical, Islamic banking that has perhaps been the most dramatic aspect of Islam's involvement with banking and investment.

Traditional Islamic economics were dominated by an injunction in the Qur'an not to charge interest on loans of money. Qadi Abu Bakr, a 12th-century Muslim teacher, describes usury as 'any unjust increment between the value of the goods given and the value of the goods received'. Muslims say that true trade is based on justice, that is, the exchange of equal for equal. Usurious trade is based on injustice, which is asking for more and giving less or nothing in return. Islam is not against the making of profit, but, according to Islamic law, this profit can only be gained justly if the person who invests risks both success and failure. The basis of the argument against interest is that the investor takes virtually no risk and does not contribute to the exchange.

Through interaction with Western banking systems, however, interest charges have found their way into Islamic life. Within almost all Islamic states, and amongst many Muslim groups, there has grown a campaign to create a non-interest-based financial system. This has been remarkably successful, and many Islamic countries, and many non-Muslim countries too, now have Islamic banking based on Shari'ah law. Indeed, so powerful has this movement become, almost all mainstream secular banks now offer Islamic banking as part of their range of services. This movement only began in earnest in 1974.

Early Judaism and Christianity also banned the charging of interest, or usury. Judaism was forced to relinquish this stand because of the precarious existence of Jews within first the Roman Empire and later within various Christian countries and empires. The Christian Church forbade usury during the Middle Ages. In the mid-16th century, however, Protestant Calvinist reformers gave usury official approval. Increasingly, Christian groups are questioning the use of interest, and many groups see interest as one of the major causes of world poverty – leading to campaigns for debt relief targeted at institutions such as the International Monetary Fund and the World Bank.

In part inspired by this, and in part concerned about injustice in trade and economics, Christian and Jewish groups have for many years sought to influence big business through their role as shareholders. Until the beginning of the 21st century, this largely took the form of withdrawing, or threatening to withdraw, funds from companies whose environmental and social policies were considered to be unjust or detrimental. Increasingly, this negative path has been joined by a positive path – seeking out and investing in companies that practise good social justice policies and are environmentally minded. Around the world, a number of organizations now exist to assist religious investors to invest ethically, ranging from specific country-based groups such as the Interfaith Center for Corporate Investment in the USA, with over 250 religious groups involved and a total asset base of over $110 billion, to the International Interfaith Investment Group (3iG), which brings together many different religions in a number of countries and advises them on ethical investment policy.

Sources

Interfaith Center on Corporate Responsibility www.iccr.org
3iG, International Interfaith Investment Group www.3ignet.org
Alliance of Religions and Conservation, ARC www.arcworld.org

The Future

(pages 78–79)

Looking into the future, even for an atlas of religion, is a risky business. But as everyone, including big business and the military, engages in the 'art and science of futurism', there seems no reason not to make an attempt to predict long-term trends.

Secularism is likely to continue to spread in Europe, especially as a result of countries joining the EU. In many parts of Western Europe, in Australasia and Canada, the growth of secularism will be largely at the cost of institutional Christianity, as expressed in the major denominations, but it will also affect minority religious communities, such as Jews, Muslims and Hindus, as constraints on family ties weaken. Although a growth of religion has been set in motion by the collapse of communism, well-established secularism will simply continue in some states of the former USSR and Eastern Europe. This will have an added component of increased pluralism, as new religions arrive and as young people explore beyond traditional boundaries of both Christianity and secularism. In Poland, in particular, there is likely to be an increase in secularism and religious pluralism. This will partially be a reaction to the powerful hold of the Catholic Church and tensions between State and Church, which are already evident, but also as a result of so many young Poles experiencing the pluralism and secularism of cities such as London or Paris.

There will be a greater willingness to belong to small, intimate and non-hierarchical religious groups, and a tendency to reject more structured and authoritarian organizations. This is already happening in many Christian countries, such as the USA, the UK and France. This tendency gives the traditional religions an opportunity to re-express themselves, with the promise of regained strength.

In the emerging states of Central Asia, the fall of communism has led to the rise of Islam. Some are looking for an Iranian-style strict Islamic state, based on Shari'ah, or have begun to take the path of the Taliban and Al-Qaeda in a rejection of anything deemed non-Islamic. Other states are looking for a more Turkish, secular model, in which religion nevertheless has an honoured place. The outcome will be significant. Iranian-style states and Taliban-influenced movements will increase religious tension and give impetus to a missionary drive in bordering states such as Turkey, China and Russia. States that follow a Turkish model will have a stronger influence in relation to trade and commerce.

The role of Islam in attacking pluralism in many countries will increase, and religious tensions will rise as a result, especially in Western countries that are labelled by such extreme groups as 'Crusader states'. The reaction to such extremism by other religions and by secular groups and governments will only fuel the sense of martyrdom and increase religious tensions. However, signs of a swing back to moderate Islam can be detected, and the next decade or more will be a deciding one for the older virtues of tolerance.

Buddhism will continue to recover in China and Mongolia and other communist or nationalist areas of Asia. However, as has been seen in the case of Daoism in China, the rebuilding of tradition after terrible loss is often complicated, and elements of the tradition need to be re-conceived. Such changes are even being fuelled by the arrival of Western Buddhism in Asia.

The appeal of a moral, ethical and religious code, which makes little or no accommodation for anti- or non-religious standards, is highly attractive to those confused and disturbed by social, political and religious change. While this return to fundamentals may be interpreted as a backward movement, it also has revolutionary potential. As well as Islamic fundamentalists, others – such as the nationalist, Orthodox, right-wing groups in Russia and the extreme nationalist Hinduism in India – are going down this fundamentalist path.

Fundamentalism posits perhaps the only serious challenge to the growth of rationalistic, consumerist, secular society. Given the appeal of fundamentalism to most of the developing world (whether whole states, as in Central Asia, or poor communities, such as the urban poor of Central America and the black urban poor of the USA), those preaching fundamentalism within the different religions have begun to take social and political issues seriously. For example, the Hezbollah in Lebanon are in the government and run the best welfare systems in Lebanon. The potential for fundamentalism to become a major voice of the oppressed could radically alter not just religious, but political, maps around the world.

The historic Christian communities of Western Asia (the Near and Middle East) will continue to emigrate as Islamic militancy grows. Highly localized expressions of ancient Christianity are now scattered over 50 to 70 countries worldwide. Also the result of migration, the rise of significant Islamic communities in Europe means greater religious diversity, and appears to be leading to greater tensions between Muslims and the host communities.

Christianity and Islam will continue to spread across the world, especially in Asia and Africa, and largely at the expense of traditional beliefs. However, the encounter will produce more hybrid groups, which fuse elements of Christianity or Islam with traditional beliefs.

Traditional beliefs may also lose ground in China. With the gradual removal of more strict Marxist-Leninist-Maoist thought, the Chinese are not only turning back to traditional ways but also towards Christianity and a revitalized Chinese Buddhism.

All the major religions are now accessible across most of the world. As yet, most people still follow the main historic religion of their area. However, the rise of Christianity and Islam is continuing so rapidly that pluralism may, in some states, replace the traditional notion of a majority religion. South Korea, for example, is shifting away from its historic religion (in this case Buddhism) and into being a majority Christian state. The indigenous Christian churches of Africa, radical forms of Judaism in the USA, new versions of Buddhism in the West – all these are arising in places far removed from their historic centres. Religion only survives when it is able to change. The rapidity of change at present, and the widespread and diverse nature of many religions, makes it difficult to say what will eventually emerge.

Sources

BBC: http://www.bbc.co.uk/religion
Other News http://soros.c.topica.com/maafi0rabujyDb4Vik4b
Medley Partners www.medleypartners.com

Origins

(pages 82–83)

The historic roots of the world's major religions and their divisions lie in a swathe across the 'old' world: from Europe, across the Middle East, India and up into China and Japan. These are the roots of Buddhism in Nepal, north India, Tibet, South-East Asia, China and Japan; of Christianity in Europe and Western Asia; of Hinduism in India; Islam in Western Asia and North Africa; and Judaism in the Middle East and Eastern Europe. Over the centuries these have been the centres of the great faiths as well as the source of their most significant divisions, all of which still exist today.

The inclusion of the Church of Jesus Christ of the Latter-day Saints marks the beginning of what may well prove to be a shift in religious geography. While claiming to be part of Christianity, the Mormons – as they are known – are a separate and distinct religious tradition. Increasingly, new religious movements are emerging in countries far removed from the traditional centres. Over time it is possible that these will come to have a place on such a map.

Sources

Geoffrey Barraclough (ed) *Times Atlas of World History* (4th edition), Times Books, 1995

Martin Palmer (ed) *The Times World Religions: A Comprehensive Guide to the Religions of the World*, HarperCollins, 2004

Holy Natural

(pages 84–85)

In many cultures, natural beauty has been celebrated as being in some special sense divine. Hinduism, Daoism and traditional beliefs, for instance, consider all of nature to be in some way sacred. Certain sites are especially revered, and specific deities or stories may come to be associated with them, but usually their holiness stems from their sheer awe-inspiring presence. This map focuses only on those sites of significance within the major world religions. It would have been impossible, on the scale available, to include the vast number of sites associated with traditional beliefs, although Australian aborigines and the Kikuyu of Kenya, to name just two examples, have many sacred natural sites.

Mountains are the most common natural feature to be revered as holy. Often rising dramatically from the plains, as do Mount Ararat and Mount Sinai, it is easy to see why they have been special places since ancient times. Remoteness and the hardship experienced by those who try to live on them are also a part of their appeal. This is especially true of the major holy mountains of China. Chinese poets, painters and mystics have always found mountains to have particular significance. The Chinese character for a sage or immortal combines the characters for a person and a mountain.

The Celts, both Christian and pre-Christian, found water and islands especially mystical, and islands still exert special attraction for the religious mind, offering solitude and peace. They may also feature a mountain (the interaction of water and mountain has extra power), be difficult to reach and cut off from the easier ways of the mainland. The flowing of rivers and the cycle of rainfall, river and evaporation may be readily linked with the cycle of life, and this is a feature of many Hindu sacred places.

Belief in sacred natural places helps to preserve our natural environment as a whole. Farming and killing are not usually allowed in holy places and, in these areas, wildlife and habitats are especially protected. In recent years this has become widely acknowledged, to the extent that a study of the link was undertaken jointly by the World Wide Fund for Nature (WWF) and the Alliance of Religions and Conservation (ARC). This showed that many national parks and wildernesses were also sacred sites, and in fact many owed their very survival to their religious significance. As the major faiths recognize the importance of the environment, and the environmental movement recognizes the role of religions, holy natural sites are taking on a new significance. WWF and ARC are actively engaged in moves to have the term 'sacred site' officially recognized as a term of international environmental protection and significance, thus helping to preserve both the environment and religious sites.

Source

Nigel Dudley, Liza Higgins-Zogib, Stephanie Mansourian et al, *Beyond Belief: Linking faiths and protected areas to support biodiversity conservation*. A research report by World Wide Fund for Nature, Equilibrium and the Alliance of Religions and Conservation (ARC), WWF, 2005

Index